International Series in Operations Research & Management Science

Volume 171

For further volumes
www.springer.com/series/6161

Vivek S. Borkar • Vladimir Ejov • Jerzy A. Filar
Giang T. Nguyen

Hamiltonian Cycle Problem and Markov Chains

Springer

Professor Vivek S. Borkar
Department of Electrical Engineering,
IIT, Powai, Mumbai 400076, India

Associate Professor Vladimir Ejov
Flinders Mathematical Sciences Laboratory
School of Computer Science, Engineering
and Mathematics
Flinders University
Bedford Park, SA 5042, Australia

Professor Jerzy A. Filar
Flinders Mathematical Sciences Laboratory
School of Computer Science, Engineering
and Mathematics
Flinders University
Bedford Park, SA 5042, Australia

Dr Giang T. Nguyen
Département d'Informatique
Université libre de Bruxelles
CP 212, Boulevard du Triomphe, 2, 1050
Brussels, Belgium

ISSN 0884-8289
ISBN 978-1-4614-3231-9 ISBN 978-1-4614-3232-6 (eBook)
DOI 10.1007/978-1-4614-3232-6
Springer New York Dordrecht Heidelberg London

Library of Congress Control Number: 2012933371

Printed on acid-free paper

Springer is part of Springer Science+Business Media (www.springer.com)

To the new generation of mathematicians from developing countries who will, undoubtedly, have a profound influence on the discipline in the coming decades.

Preface

Graphs and networks have been studied extensively in recent decades by mathematicians, computer scientists, engineers, operations researchers as well as physicists, biologists, chemists, and even linguists and sociologists. Their two key elements, vertices and edges, are extremely useful as representations of a wide spectrum of phenomena ranging from transportation networks, through topology of atoms to social networks. Furthermore, many problems modelled with graphs and networks naturally lend themselves to algorithmic analysis and ultimate solutions with the help of modern high-speed computers. The shortest path, maximal spanning tree and max-flow/min-cut problems are just three examples out of a large collection of well-solved important problems.

Nonetheless, there is also a large collection of graph theoretic and network optimisation problems that are fundamentally difficult in the sense of belonging to the very challenging computational complexity classes such as the NP-complete and NP-hard classes. Indeed, the famous Hamiltonian cycle problem (HCP) is known to be NP-complete. The now extensive body of research into the HCP was, perhaps, stimulated by investigations of interesting instances of that problem by great mathematicians such as Euler in the 18th and Hamilton in the 19th century, respectively.

The essence of the Hamiltonian cycle problem is contained in the following—deceptively simple—single sentence statement:

Given a graph, find a cycle that passes through every single vertex exactly once, or determine that this cannot be achieved.

Such a cycle is called a Hamiltonian cycle. The HCP has become a challenge that attracts mathematical minds both in its own right and because of its close relationship to the famous travelling salesman problem (TSP), that calls for the identification of a Hamiltonian cycle with the lowest cost possible in a graph where every edge has a known cost associated with "travelling" along that edge. An efficient solution of the TSP would have an enormous impact in operations research, optimisation and computer science. However, from a mathematical perspective, the underlying difficulty of the TSP is, perhaps, hidden in the Hamiltonian cycle problem. Hence, in this monograph, we focus on the Hamiltonian cycle problem.

Arguably, the inherent difficulty of many problems in graph theory and combinatorial optimisation stems, precisely, from the discrete nature of the domains in which these problems are posed. Consequently, this monograph is devoted to a line of research that maps such problems into convex domains where continuum analysis can be easily carried out. This convexification of domains is achieved by assigning probabilistic interpretation to the key elements of the original problems even though these problems are deterministic.

While there are other instances of similar ideas being exploited elsewhere, our approach builds on the innovation introduced in Filar and Krass [49] where the Hamiltonian cycle problem and the travelling salesman problem are embedded in a structured singularly perturbed Markov decision process (MDP). The unifying idea of [49] is to interpret subgraphs traced out by deterministic policies (including Hamiltonian cycles, if any) as extreme points of a convex polyhedron in a space filled with randomised policies.

This approach was continued by Chen and Filar [22] and, independently, by Feinberg and Shwartz [46] and Feinberg [44]. Further results were obtained by Filar and Liu [51], Andramonov *et al.* [4], Filar and Lasserre [50], Ejov *et al.* [30]–[38] and Borkar *et al.* [17]–[18]. In addition, three recent (but not readily accessible) PhD theses by Nguyen [81], Haythorpe [62] and Eshragh [41] contain some of the most recent results. Thus, there is now an active group of researchers in various countries interested in this approach to discrete problems. Majority of these contributions focused on the classical Hamiltonian cycle problem, but in principle many of the techniques used could be adapted to other problems of discrete mathematics (as, indeed, was done by Feinberg [45]).

To indicate the flavour of the results reported in the present monograph, consider a key observation that led to the recent results presented in Borkar *et al.* [17] and [18]: the "natural" convex domain where Hamiltonian cycles should be sought is the set of doubly stochastic matrices induced by a given graph. This observation is nearly obvious, once we recall the famous Birkhoff-von Neumann theorem, which states that the set of all $N \times N$ doubly stochastic matrices is the convex hull of permutation matrices. Of course, in searching for a Hamiltonian cycle of a given graph, we need to restrict ourselves to the convex hull of only those permutation matrices that correspond to subgraphs of that graph. Results in Chapter 3 (based on Borkar *et al.* [17] and [18]) imply, that after a suitable perturbation and defining the random variable τ_1 to be the first hitting time of the home vertex 1 (after time 0), the Hamiltonian cycle problem essentially reduces to "merely" minimising the variance-like functional $\mathbb{E}[(\tau_1 - N)^2]$ over the space of doubly stochastic matrices. This probabilistic, almost statistical, interpretation enables us to exploit a wide range of both analytical and algorithmic tools on the HCP.

More generally, this monograph summarises results of both theoretical and algorithmic investigations. The theoretical aim of this line of research is to explain the essential difficulty of the Hamiltonian cycle problem in analytic terms such as a measure of variability, or the size of a gap between certain optimisation problems, or by the nature of certain singularities. The algorithmic aim of the approach is to construct either exact or heuristic methods to obtain numerical solutions of the HCP. It is based on the belief that some classical "static" optimisation problems can be well analysed by embedding them in suitably constructed Markov decision processes.

In our setting, the theoretical and algorithmic aims are not separate. Indeed, results on one aim seem to influence progress on the other. For instance, the optimisation algorithms presented in Chapters 7 and 8 follow directly from the theoretical developments presented in Chapters 3–5 and have identified difficulties that some of the theoretical developments reported in Chapters 6, 9 and 10 are trying to resolve.

The general approach constitutes one of the few instances where probabilistic, continuous optimisation and dynamic control methods are combined to deal with a hard problem of discrete mathematics. Arguably, simulated annealing could be seen as a precursor of this approach. However, it should be mentioned that relationships between Markov chains and graphs are also of recent interest to other researchers, notably Aldous and Fill [2] and Hunter [67].

Next we shall, briefly, differentiate between our approach and some of the best known, well established, approaches to the HCP. We first note that the present line of research is essentially different from that adopted in the study of *random graphs*, where an underlying random mechanism is used to generate a graph (see, for example, Karp's seminal paper [69]). In our approach, the graph to be studied is given and fixed but a *controller* can choose edges according to a probability distribution, and with a small probability (due to a perturbation) an edge may take you to a vertex. Random graphs have played an important role in the study of Hamiltonicity, a striking result to quote is that of Robinson and Wormald [92] who show that *with high probability* k-regular graphs are Hamiltonian, for $k \geq 3$.

Typical general purpose heuristic algorithms can perhaps be classified—we cite only few representative papers—as *rotational transformation* algorithms (Posa [86]), *cycle extension* algorithms (Bollobas *et al.* [13]), *long path* algorithms (Kocay and Li [71]), *low degree vertices* algorithms (Broder *et al.* [20] and Brunacci [21]), *multipath search* or *pruning* algorithms (Christofides [23]). Of course, much research has been done on algorithms for finding a Hamiltonian cycle on various restricted graph classes (see, for example, Parberry [84]). Clearly, algorithms designed for particular classes of graphs tend to outperform the best general purpose algorithms when applied to graphs from these classes.

In the operations research and optimisation communities, many of the successful, now classical, approaches to the HCP and TSP focus on solving a linear programming relaxation followed by heuristics that prevent the formation of sub-cycles (see, for example, Lawler *et al.* [76]). In the present approach, we embed a given graph in a singularly perturbed MDP in such a way that we can identify Hamiltonian cycles with irreducible Markov chains and sub-cycles with non-exhaustive ergodic classes. This permits a search for a Hamiltonian cycle in either (i) the policy space of an MDP, or (ii) the space of the occupational measures of the MDP that is a polytope with a non-empty interior. In both cases, the original discrete optimisation problem is converted to a continuous one. The Branch and Fix, Wedged-MIP and Cross-Entropy heuristics reported in Chapters 7 and 8 can be seen as belonging to (ii), as they all exploit properties of the spaces of occupational measures. They are performing competitively with alternative—general purpose—algorithms on various test problems including the Knight's Tour problem on chessboards of the size up to 32×32. The Interior Point heuristic discussed in Chapter 8 belongs to (i) and should be properly seen as being still under development. However, it opens up promising opportunities for a lot of further research, as it exploits numerically attractive algebraic factorisation properties of irreducible generator matrices of Markov chains.

Indian Institute of Technology, India, *Vivek S. Borkar*
Flinders University, Australia, *Vladimir Ejov*
Flinders University, Australia, *Jerzy A. Filar*
Université Libre de Bruxelles, Belgium, *Giang T. Nguyen*

Acknowledgements

This survey summarises a line of research initiated in 1985 when the third author proposed a stochastic approach to the Hamiltonian cycle problem as a thesis topic for his former PhD student, Dmitry Krass. Since then investigations continued, albeit with numerous interruptions, but with important contributions from many collaborators, colleagues and research students. The authors gratefully acknowledge invaluable assistance and insights gained from collaboration or merely from discussions with Michael Andramonov, Konstantine Avrachenkov, Pouya Baniasadi, Ming Chen, Eric Denardo, Eugene Feinberg, Jacek Gondzio, Arie Hordijk, Lodewijk Kallenberg, Jean-Bernard Lasserre, Ke Liu, Nelly Litvak, Stephen Lucas, Wayne Lobb, Walter Murray, Kien Ming Ng, Jessica Nelson, Minh Nguyen, Panos Pardalos, Ulrich Rieder, Serguei Rossomakhine, the late Alex Rubinov, Floske Spieksma, Peter G. Taylor, Jane Thredgold and Peter Zograf.

Fellow collaborators Ali Eshragh Jahromi and Michael Haythorpe deserve a special thanks and an apology. Thanks, because they played a central role in the development of some of the results, and an apology for infecting them with the "Hamiltonicity virus" that makes it hard to let go of this problem. They also proof-read significant sections of the book manuscript and their youthful energy and enthusiasm provided great support. The last author is grateful for the support of her postdoctoral supervisor, Guy Latouche, during the completion of this book. Last, but not least, much of this research has been partially supported by grants from the Australian Research Council (most recently, by the Discovery grants DP0666632 and DP0984470) and the Ministère de la Communauté française de Belgique (ARC grant AUWB-08/13ULB 5).

Contents

Part I
Motivating Phenomena

Chapter 1
Illustrative Graphs

1.1 The Graph That Started It All

Sir William Rowan Hamilton (1805–1865) was a famous mathematician and physicist, well-known for his vast contribution in various fields such as optics, classical mechanics and algebra. However, it is a lesser known fact that Sir William was the inventor of a commercial game, the mathematically generalised version of which later became one of the most difficult graph theoretic problems. Towards the end of his life, in 1857, Sir William Rowan Hamilton designed a game called *Icosian*. Its name derived from a Greek word *Icosa*, meaning twenty, the Icosian Game featured twenty connected cities, each represented by a hole on a wooden pegboard. Deceptively simple, the player was to visit every city exactly once and return to where he or she started. If we represent each city in this game by a vertex and each connection between two cities by an edge, then the resulting map of cities in the Icosian game is the Dodecahedron graph (Figure 1.1).

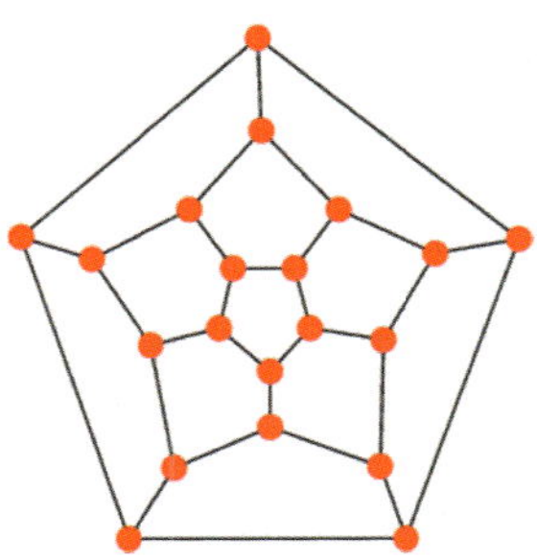

Fig. 1.1: The dodecahedral graph [102], which represents the city map of Icosian game, is the planar projection of a dodecahedron (a polyhedron with twelve faces)

A mathematically generalised version of the Icosian game, the *Hamiltonian cycle problem* (HCP) can be succinctly stated as:

Given a graph, find a cycle that passes through every single vertex exactly once, or determine that this cannot be achieved.

Such a cycle is called a *Hamiltonian cycle* (HC). A graph is said to be *Hamiltonian* if it possesses at least one Hamiltonian cycle, and is otherwise *non-Hamiltonian*. There might be, and usually are, multiple Hamiltonian cycles in a Hamiltonian graph. The dodecahedron graph is Hamiltonian and has multiple Hamiltonian cycles. The property of a graph possessing a Hamiltonian cycle is called *Hamiltonicity*.

In 1859, Sir William sold the game to an Irish toy manufacturer for 25 British pounds, approximately equivalent to today's 2770 US dollars. The Icosian game was later commercially distributed as the *Traveller's Dodecahedron*. In 2000, the Clay Mathematics Institute announced their list of seven *Prize Problems*, and offered a one million US dollar prize for a solution to each problem. One listed problem is the long-standing question on the relationship between two complexity classes P and NP (see Cook [26] for the formal problem description), an answer to which can be found by determining whether there exists a polynomial-time algorithm to solve the Hamiltonian cycle problem.

In the even better known travelling salesman problem, where we assign a cost for each edge in a given graph, the objective is essentially to determine *which* Hamiltonian cycle on the graph is the most cost-efficient. Thus, the Hamiltonian cycle problem is a special case of the travelling salesman problem, and both are computationally difficult to solve. An efficient solution to the Hamiltonian cycle problem would help solve the travelling salesman problem effectively, and therefore would have a great impact in various fields such as computer science, operations research and cryptology.

1.2 A Sample of Distinctive Graphs

A simple indication of the complexity of the Hamiltonian cycle problem is, that it is not easy to determine whether or not a graph is Hamiltonian by inspection even for small-sized problems. For example, it might take several minutes for a person to determine whether the famous 10-vertex Petersen graph (Figure 1.2) is Hamiltonian. A larger, but still reasonably small in terms of computer science instances, problem that has challenged many HCP algorithms is the 96-vertex Horton graph (Figure 1.3). In both graphs, there is no Hamiltonian cycle.

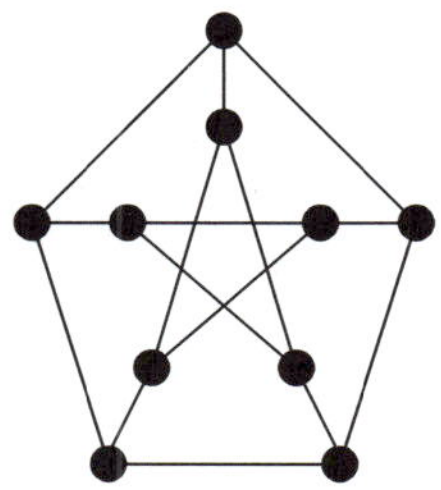

Fig. 1.2: Petersen graph

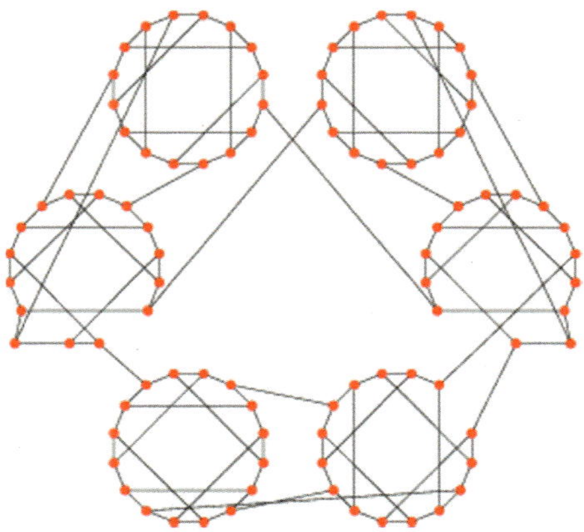

Fig. 1.3: Horton graph [103]

The dodecahedral graph, the Petersen graph and the Horton graph are all *cubic* graphs. A graph is *cubic* or *3-regular* if every vertex in the graph is connected to exactly three other vertices. In general, a graph is *k-regular* if every vertex in the graph is connected to exactly k other vertices. Despite this regularity constraint which seemingly simplifies things, the Hamiltonian cycle problem when restricted to cubic graphs retains its full complexity and hence remains NP-complete [53]. Consequently, cubic graphs are one of the simplest classes of graphs frequently chosen for analysis, in regards not only to the Hamiltonian cycle problem but also to many other graph theoretic problems [57].

In Figure 1.4, we list all 19 connected 10-vertex cubic graphs, 17 of which are Hamiltonian and two are non-Hamiltonian, including the Petersen graph. We enumerate these graphs using the graph-generating GENREG software (Meringer [79]). For the Hamiltonian 10-vertex cubic graphs (numbered 2 to 18), all vertices are drawn in a circle to highlight the Hamiltonicity of these graphs. Indeed, each of these Hamiltonian graphs has at least one Hamiltonian cycle that is the circle circumscribing the graph. However, both non-Hamiltonian graphs (numbered 1 and 19) are drawn with a different vertex arrangement. This diagrammatic version of the first graph highlights a *bridge* connecting two subgraphs, consequently indicating the lack of Hamiltonian cycles in the graph.

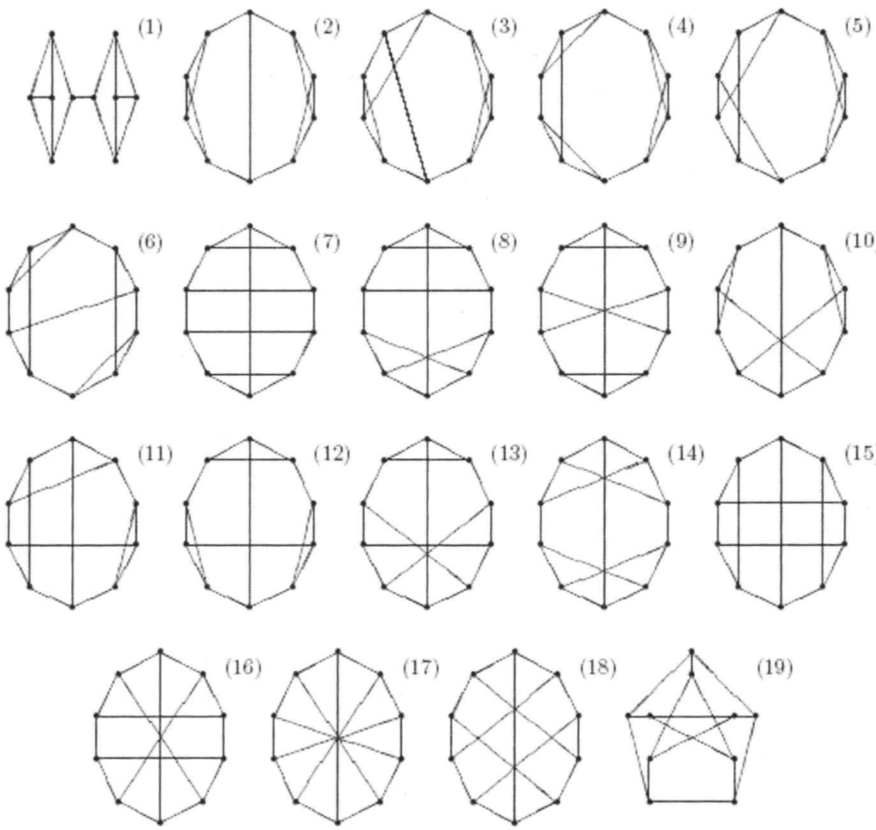

Fig. 1.4: All 19 connected 10-vertex cubic graphs

For any given graph, there are many different graphical representations. A *planar* graph is a graph that can be drawn in the plane in such a way that its edges intersect only at the vertices. A *nonplanar* graph is a graph that is not planar. Garey *et al.* [53] show that, even when restricted to planar cubic graphs, the Hamiltonian cycle problem is NP-complete.

In addition to the quest for finding an efficient algorithm to solve the Hamiltonian cycle problem for generic graphs, a lot of focus is on determining whether classes of graphs with certain properties are always Hamiltonian. Before stating a famous conjecture on a particular class of graphs, we need to introduce a few definitions. A graph is *bipartite* if its set of vertices can be divided in two disjoint subsets such that no pair of vertices in the same subset are connected to each other. In general, a graph is *k-partite* if its set of vertices can be divided into k disjoint subsets such that no two pair of vertices in same subset are connected to each other. A graph is *k-connected* if it remains connected after we remove any set of $k-1$ vertices from the graph. Of course, a cubic graph can be at most 3-connected, since removing the three vertices adjacent to any vertex isolates it.

Among the aforementioned graphs, the dodecahedral graph is planar and 3-connected; the Petersen graph is nonplanar, 3-partite and 3-connected; the Horton graph is nonplanar, bipartite and 3-connected. Out of 19 connected 10-vertex cubic graphs in Figure 1.4, only the graph numbered 1 is 1-connected, as the removal of either vertex at the endpoints of the bridge disconnects the graph, and this graph is non-Hamiltonian. In fact, it was shown almost a century ago that every 1-connected graph is non-Hamiltonian [72]. It is easy to verify that 1-connected graphs can be detected in polynomial time.

A famous, long-standing and still open conjecture on Hamiltonicity of graphs is the following conjecture, which relates Hamiltonicity to connectivity, regularity and planarity.

Conjecture 1.1. ***Barnette's Conjecture*** [100]. *Every 3-connected bipartite cubic planar graph is Hamiltonian.*

1.3 Co-spectral Graphs

A common way to represent a graph with N vertices is using an $N \times N$ *adjacency matrix*, of which the (i,j)th entry is 1 if edge (i,j) is present in the graph and 0 otherwise. Two graphs are *co-spectral* if their adjacency matrices share the same set of eigenvalues. In various texts, associated with a graph eigenvalues could also be the *Laplacian matrix* or the *normalised Laplacian matrix* of the graph (see Chapter 2 for precise definitions of these matrices).

Consider two graphs G and H, and let $V(G)$ and $V(H)$ be the sets of vertices, and $E(G)$ and $E(H)$ be the sets of edges in G and H, respectively. For brevity, we drop the dependency on graphs and simply write E and V when no confusion can arise. The graphs G and H *isomorphic* if there exists a bijection $f : V(G) \mapsto V(H)$ such that for every edge $(u,v) \in E(G)$, the edge $(f(u), f(v)) \in E(H)$. They are *non-isomorphic* otherwise. An automorphism of a graph G is an *isomorphism* of G with itself. Isomorphic graphs are co-spectral. Cvetkovic [27] shows that there exist non-isomorphic graphs that are co-spectral, and van Dam and Haemers [28] discuss the types of graphs that are uniquely determined by their spectrum. More directly related to the Hamiltonian cycle problem, Filar *et al.* [47] construct two 20-vertex cubic, isomorphic and co-spectral graphs, one of which is Hamiltonian and the other non-Hamiltonian.

This indicates that we cannot rely on the spectrum of a graph alone to determine whether the graph is Hamiltonian or not. However, the eigenvalues and consequently the determinant of the adjacency matrix of a graph still contain a lot of useful insights that could help us with determining Hamiltonicity. For example, in an N-vertex Hamiltonian graph, it is well-known that the characteristic polynomial of an adjacency matrix of a subgraph corresponding to any Hamiltonian cycle on the graph is $\lambda^N - 1$. Moreover, Ejov *et al.* [32] show that, for a given N-vertex Hamiltonian graph, any Hamiltonian cycle is equivalent to a solution of the following system of polynomial equations:

$$\begin{aligned} x_{ij}(1 - x_{ij}) &= 0 \quad \text{for all } (i,j) \in E, \\ \sum_j x_{ij} - 1 &= 0 \quad \text{for all } i \in V, \\ \sum_i x_{ij} - 1 &= 0 \quad \text{for all } j \in V, \\ \det(\lambda\mathbf{I} - \mathbf{X}) - \lambda^N + 1 &= 0, \end{aligned}$$

where $\det \mathbf{A}$ denotes the determinant of a matrix $\mathbf{A}$, $\mathbf{I}$ is an $N \times N$ identity matrix, and $\mathbf{X}$ is the *modified adjacency matrix*, defined to be the adjacency matrix with every non-zero (i,j)th entry replaced by the variable x_{ij}, for all $i, j \in V$.

In Chapter 2, we discuss a few motivating numerical observations about eigenvalues and determinants of subgraphs that lead to theoretical results presented in the subsequent chapters.

Chapter 2
Intriguing Properties

2.1 Preliminaries and Notation

In this book, all graphs are connected and undirected, unless otherwise stated. We follow the graph terminology and conventions from Harary [59], where the reader can find an excellent introduction to graph theory. Consider a graph $G = (V(G), E(G)) = (V, E)$, where V is the set of vertices of G, $|V| = N$, and E is the set of edges of G. A graph with N vertices is said to be a graph of *order* N. As the graph G is undirected, for every edge $(i, j) \in E$ there exists an opposite edge $(j, i) \in E$, where $i \neq j$. A *loop* is an edge (i, i) joining a vertex to itself. We do not consider *multi-edges*, which are distinct edges that connect the same pair of vertices, and we use the terms *edge* and *arc* interchangeably.

Subgraphs and Regularity Consider a graph G', and let $V(G')$ be the set of its vertices and $E(G')$ be the set of its edges. Then G' is a *subgraph* of G if $V(G') \subseteq V(G)$ and $E(G') \subseteq E(G)$. A subgraph G' is a *spanning subgraph* if $V(G') = V(G)$. From now on, the term *subgraph* refers to a spanning subgraph, unless otherwise stated. A vertex j is a *neighbour* of i, or vertex j is *adjacent* to i, if there exists an edge between them, that is, $(i, j) \in E(G)$. A vertex v has a *degree* d if it has d neighbours, and we write $\deg(v) = d$. A graph is *k-regular* if every vertex $i \in V$ has the same degree k, and a *cubic* graph is a 3-regular graph.

Walks, Paths and Cycles A *walk* is a sequence of vertices $(v_0, v_1, \ldots, v_n)$ where each edge $(v_i, v_{i+1}) \in E$ for $i = 0, \ldots, n-1$. A walk is said to be *closed* if $v_0 = v_n$, and *open* otherwise. A walk is a *path* if all vertices in the sequence are distinct, that is, $v_i \neq v_j$ for all $i \neq j$. A path is a *cycle* if it is closed. The *length* of a walk, a path or a cycle is the number of edges on the walk, the path or the cycle, respectively. The *girth* of a graph is the length of a shortest cycle on the graph, excluding cycles of length two. On the other hand, the *circumference* of a graph is the length of a longest cycle on the graph. The circumference of a Hamiltonian graph of order N is N, as any Hamiltonian cycle is a longest cycle of the graph.

Example 2.1 *We give examples of an open walk (Figure 2.1), an open path (Figure 2.2) and a cycle (Figure 2.3).*

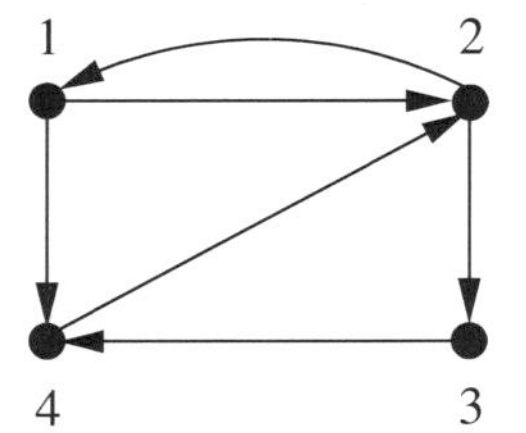

Fig. 2.1: An open walk $(1, 2, 3, 4, 2, 1, 4)$

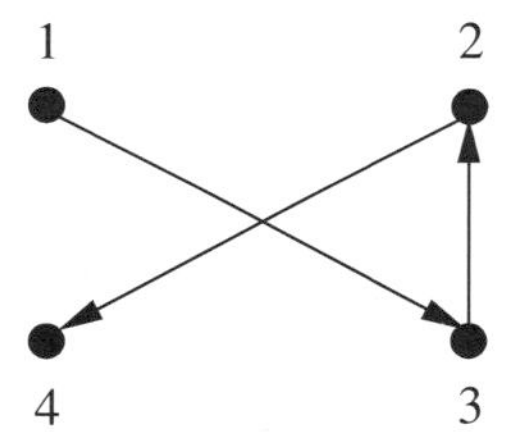

Fig. 2.2: An open path $(1, 3, 2, 4)$

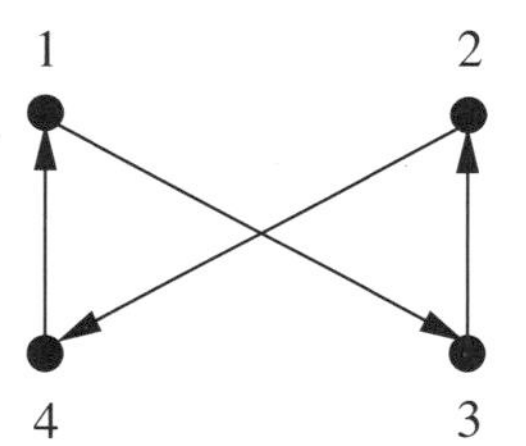

Fig. 2.3: A cycle $(1, 3, 2, 4, 1)$

Adjacency Matrices The *adjacency matrix* $\mathbf{A} = [a_{ij}]$ of a graph G has elements

$$a_{ij} = \begin{cases} 1 & \text{for } (i, j) \in E, \\ 0 & \text{otherwise.} \end{cases}$$

The adjacency matrix of an undirected graph is always symmetric. Because it is a one-to-one representation of a (labelled) graph, the adjacency matrix is one of the most investigated matrices in graph theory. We can obtain much interesting information about a given graph by studying its adjacency matrix. For instance, the number of walks of length k from vertex i to vertex j is the (i, j)th entry of the matrix $\mathbf{A}^k$ [28].

Laplacian Matrices Another well-studied matrix in graph theory is the *Laplacian matrix* $\mathbf{L} = [\ell_{ij}]$ where

$$\ell_{ij} = \begin{cases} \deg(i) & \text{if } i = j, \\ -1 & \text{if } i \neq j \text{ and } (i,j) \in E, \\ 0 & \text{otherwise.} \end{cases} \tag{2.1}$$

The degree matrix $\mathbf{D}$ of G has diagonal elements $d_{ii} = \deg(i)$ for $i = 1, \ldots, N$. It is clear that

$$\mathbf{L} = \mathbf{D} - \mathbf{A}. \tag{2.2}$$

A *normalised Laplacian matrix* is defined as $\widetilde{\mathbf{L}} = [\widetilde{\ell}_{ij}]$, where

$$\widetilde{\ell}_{ij} = \begin{cases} 1 & \text{if } i = j, \\ \dfrac{-1}{\sqrt{\deg(i)\deg(j)}} & \text{if } i \neq j \text{ and } (i,j) \in E, \\ 0 & \text{otherwise.} \end{cases} \tag{2.3}$$

Spectrum of a Graph The *spectrum of a graph* can refer to either the set of eigenvalues of the adjacency matrix, or of the Laplacian matrix, or of the normalised Laplacian matrix of the graph. For an excellent discussion of normalised Laplacian matrices, we refer the interested reader to Chung [24].

2.2 Fractal-like Structure of Graphs

In this section, we present an intriguing self-similar structure that groups cubic graphs according to the numbers of closed walks of various sizes in each graph. The procedure of obtaining this self-similar structure is: for each cubic graph G and its adjacency matrix $\mathbf{A}$, we evaluate the sample mean $\mu(\mathbf{A})$ and variance $\sigma^2(\mathbf{A})$ of the exponentials of all eigenvalues of $\frac{1}{3}\mathbf{A}$. As the eigenvalues of the adjacency matrix of a cubic graph are real and belong to the interval $[-3, 3]$, the eigenvalues of $\frac{1}{3}\mathbf{A}$ belong to the interval $[-1, 1]$. Thus, each cubic graph G corresponds to a point $(\mu(\mathbf{A}), \sigma^2(\mathbf{A}))$ in these mean-variance coordinates. We illustrate this procedure with the following example.

Example 2.2 *Consider the labelled Petersen graph.*

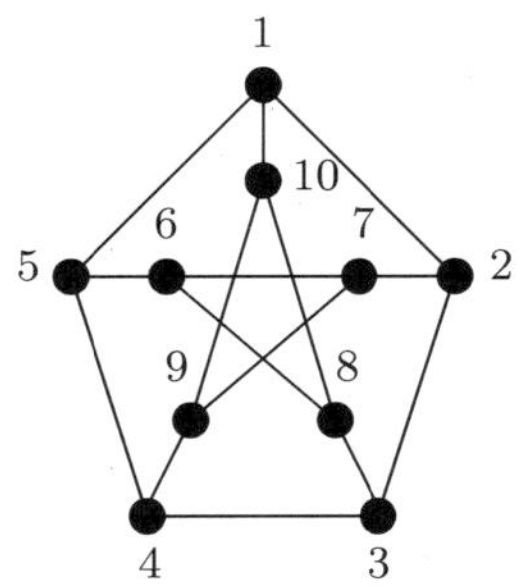

Fig. 2.4: Labelled Petersen graph

Its adjacency matrix $\boldsymbol{A}$ *is given by*

$$\mathbf{A} = \begin{bmatrix} \cdot & 1 & \cdot & \cdot & 1 & \cdot & \cdot & \cdot & \cdot & 1 \\ 1 & \cdot & 1 & \cdot & \cdot & \cdot & 1 & \cdot & \cdot & \cdot \\ \cdot & 1 & \cdot & 1 & \cdot & \cdot & \cdot & 1 & \cdot & \cdot \\ \cdot & \cdot & 1 & \cdot & 1 & \cdot & \cdot & \cdot & 1 & \cdot \\ 1 & \cdot & \cdot & 1 & \cdot & 1 & \cdot & \cdot & \cdot & \cdot \\ \cdot & \cdot & \cdot & \cdot & 1 & \cdot & 1 & 1 & \cdot & \cdot \\ \cdot & 1 & \cdot & \cdot & \cdot & 1 & \cdot & \cdot & 1 & \cdot \\ \cdot & \cdot & 1 & \cdot & \cdot & 1 & \cdot & \cdot & \cdot & 1 \\ \cdot & \cdot & \cdot & 1 & \cdot & \cdot & 1 & \cdot & \cdot & 1 \\ 1 & \cdot & \cdot & \cdot & \cdot & \cdot & \cdot & 1 & 1 & \cdot \end{bmatrix}, \tag{2.4}$$

where '·' denotes 0*. The eigenvalues of* $1/3\mathbf{A}$ *are* $1, -2/3, -2/3, -2/3, -2/3, 1/3, 1/3, 1/3, 1/3, 1/3$*. Consequently,*

$$\mu(\mathbf{A}) = \frac{1}{10}(e^1 + 4e^{-2/3} + 5e^{1/3}) = 1.1750, \tag{2.5}$$

and

$$\sigma^2(\mathbf{A}) = 0.4376, \tag{2.6}$$

and we obtain the point $(0.1750, 0.4376)$ *for the Petersen graph.*

Continuing the procedure of obtaining the self-similar structure, we then plot the mean-variance coordinates for all cubic graphs of the same order. We identify a very interesting behavior for N taking values from 10 to 18. For example, the structures for $N = 14$ and $N = 16$ appear as in, respectively, Figures 2.5 and 2.6.

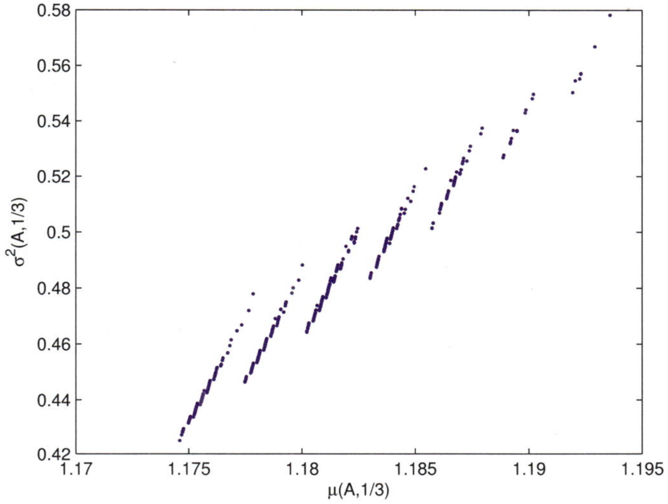

Fig. 2.5: Mean-variance plot for cubic graphs of order 14

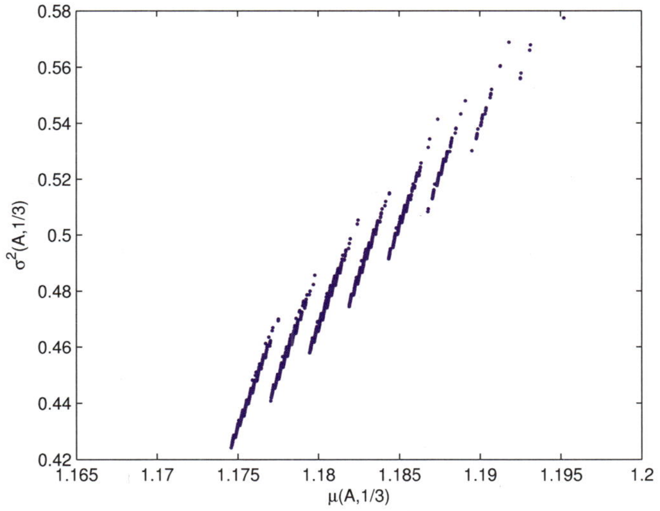

Fig. 2.6: Mean-variance plot for cubic graphs of order 16

In these two figures, it is evident that the resulting scattergram in the mean-variance coordinates consists of thread-like clusters with similar slopes of and distances between consecutive clusters. Moreover, the value of variance at the bottom of each segment is strictly increasing from left to right in the plot. Ejov *et al.* [33] use the term *multifilar* to refer to this thread-like structure, with each approximately linear cluster being called a *filar*. The authors make an important observation that the overall structure is self-similar. In particular, zooming in on each of these filars shows us similar but smaller sub-filars that are also made up of approximately straight and parallel segments, shifted gradually from left to right. We illustrate this self-similarity in Figures 2.7 and 2.8, by showing plots of two successive enlargements of the first filar of Figure 2.5.

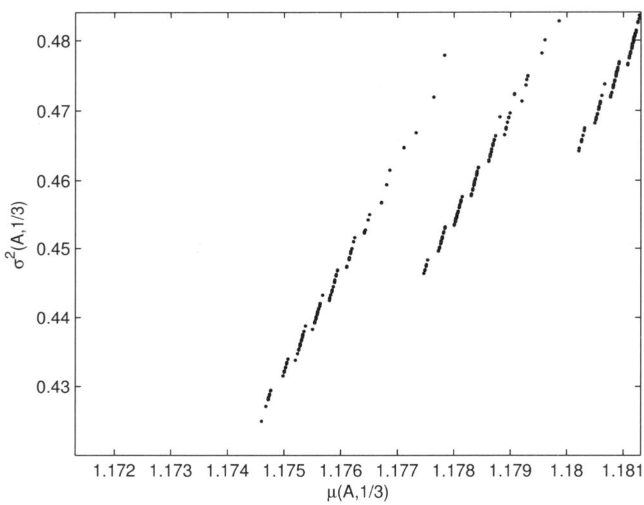

Fig. 2.7: Zooming in on the first, from the left, filar in Figure 2.5—first level

Using a form of Ihara-Selberg trace formula derived in Mnev [80], Ejov *et al.* [33] explain the filar memberships for each graph. In the overall clustering, all graphs belonging to each filar have the same number of triangles (cycles of length three) and these numbers strictly increase from the left most filar to the right most, starting from zero. In the first level of zooming-in, all graphs in a particular sub-filar have the same number of quadrangles (cycles of length four) while the number of triangles over all these sub-filars is fixed. This pattern repeats itself, with each higher level of zooming-in corresponding to a larger cycle size.

Consider another frequently used matrix function in the spectral theory of

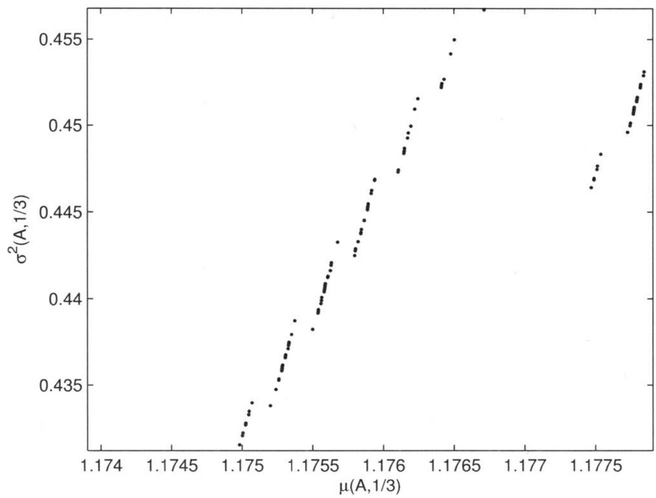

Fig. 2.8: Zooming in on the first, from the left, filar in Figure 2.5—second level

linear operators: the resolvent of $t\mathbf{A}$ for $t \in (0, 1/3)$. It appears that we can reproduce the self-similar phenomenon described above in the mean-variance coordinates with different slopes of and distances between segments [36]. In Chapter 9, we use a modification of the Ihara-Selberg trace formula [80] to justify the multifilar structure of the observed plots and to estimate the slopes of and distances between filars, consistently with numerical evidence.

It is worth noting that in the aforementioned self-similar phenomenon, non-Hamiltonian graphs seem to be separated in two groups. The first group contains *easy non-Hamiltonian* graphs that seem to be located at the tops of (the most zoomed in) sub-filars. We call a non-Hamiltonian graph *easy* if it contains one or more *bridges*. A *bridge* is an edge the removal of which disconnects the graph. A *bridge graph* is an easy non-Hamiltonian graph because these bridges, and consequently the bridge graphs, can be identified in polynomial time [72] (in fact, in linear time in N). We call a non-Hamiltonian graph that is not a bridge graph a *hard* non-Hamiltonian graph. In the self-replicating structure, *hard non-Hamiltonian* graphs (the second group) seem to be found at the bottom ends of (the most zoomed in) sub-filars. In these sub-filars, the Hamiltonian graphs are strictly in between these two groups of non-Hamiltonian graphs. To illustrate this observation, we plot the mean-variance coordinates for the trace of the matrix resolvent, over the set of all cubic graphs of order 14, in Figure 2.9, where dots represent Hamiltonian graphs and crosses represent non-Hamiltonian graphs.

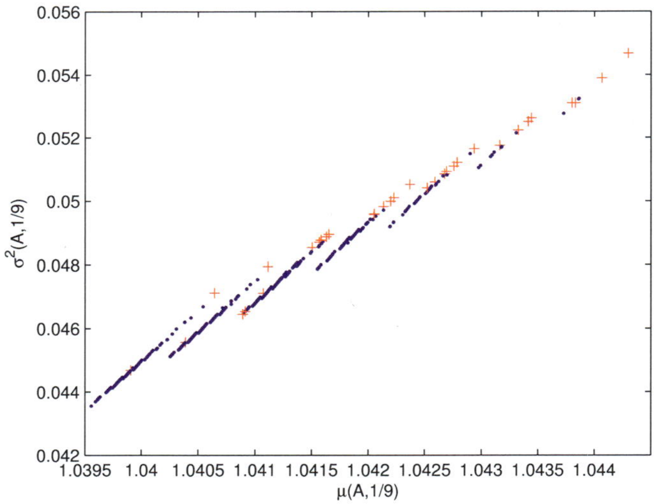

Fig. 2.9: Mean-Variance plot for cubic graphs of order 14. Here, the dots represent Hamiltonian graphs and crosses represent non-Hamiltonian graphs

We also include a zooming-in plot (see Figure 2.10) for more clarity. All crosses that can be seen clearly in this plot are either at the top or the bottom of their sub-filars.

In Chapter 9, we revisit this observation in more detail, and explore further interesting properties that arise from it.

2.3 Invariants of Graphs

A *graph invariant* is a function that maps the set of graphs to some other set, such as the set of natural numbers in such a way that certain "similar" graphs are all mapped onto the same number. Informally, we can think of a graph invariant as a numerical property associated with a graph that does not depend on the graph labelling or drawing. Examples of graph invariants include the number of vertices, the number of edges, the number of connected components, the graph spectrum, and the *chromatic number*, which is the minimum number of colours it requires to colour the vertices in a given graph in such a way that no two connected vertices share the same colour.

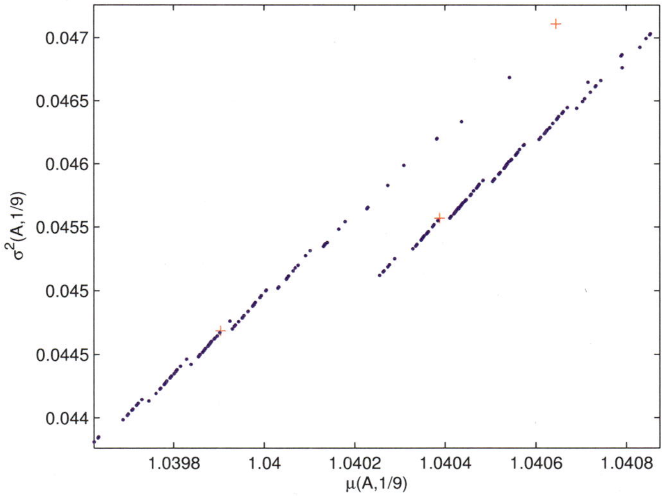

Fig. 2.10: Mean-Variance plot—zooming in. Here, the dots represent Hamiltonian graphs and crosses represent non-Hamiltonian graphs

For each graph G of order N, we introduce a set $\mathcal{F}(G)$ of matrices on G, where

$$\mathcal{F}(G) = \{\mathbf{P} \in \mathbb{R}^{N\times N} \,|\, p_{ij} = 0 \text{ if } (i,j) \notin E, \sum_{j=1}^{N} p_{ij} = 1, \\ p_{ij} \geq 0 \text{ for all } (i,j) \in E\}. \tag{2.7}$$

In Chapter 3, we show that these matrices have probabilistic interpretations. For now, we refer to $\mathcal{F}(G)$ as the set of *feasible matrices* on G, and drop the dependency on G when no confusion can arise. If p_{ij} is either 1 or 0 for all $i, j \in V$, the matrix $\mathbf{P}$ is said to be *deterministic*. Every deterministic matrix $\mathbf{P} \in \mathcal{F}(G)$ corresponds to a spanning subgraph of G that has exactly one edge coming out of each vertex.

For each $\mathbf{P} \in \mathcal{F}$, we define the matrix

$$\mathbf{W}(\mathbf{P}) = \mathbf{I} - \mathbf{P} + 1/N\mathbf{J}, \tag{2.8}$$

where $\mathbf{I}$ is the $N \times N$ identity matrix, and $\mathbf{J}$ is an $N \times N$ matrix of which every entry is unity. Next, we note that the maximum value of the determinant of $\mathbf{W}(\mathbf{P})$, over the set $\mathcal{F}$, is a graph invariant. It is in fact equal to the circumference of the graph, that is, the longest cycle of the graph. Consequently, solving the Hamiltonian cycle problem for a graph G is equivalent to maximising the determinant of $\mathbf{W}(\mathbf{P})$ over the set $\mathcal{F}(G)$. In Chapter 5, we prove the above property in a more general case, where $1/N$ is replaced by any constant $\alpha \in \mathbb{R}^+$. In the same chapter, we also show that when we apply a linear, singular perturbation to the determinant functional, Hamiltonian cycles remain the maximisers over $\mathcal{F}$.

Example 2.3 *Consider a cubic graph G of order 6, which we call the* envelope graph *(Figure 2.11). One of the only two 6-vertex cubic graphs, both of which are Hamiltonian, the envelope graph has six Hamiltonian cycles.*

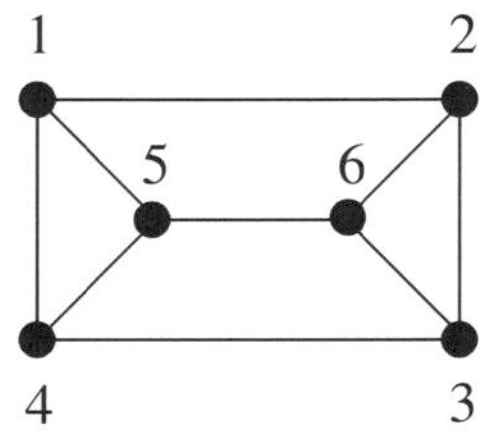

Fig. 2.11: The envelope graph

Let H_1 and H_2 be the two Hamiltonian cycles on the envelope graph, depicted in Figures 5.3 and 5.4, respectively.

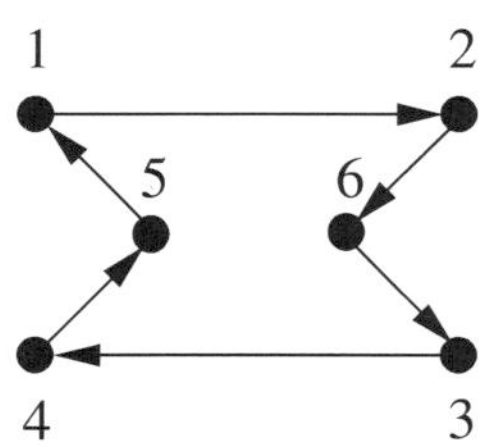

Fig. 2.12: The Hamiltonian cycle H_1

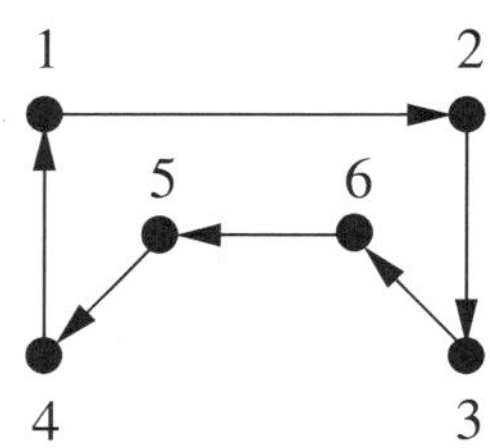

Fig. 2.13: The Hamiltonian cycle H_2

The transition matrices $\mathbf{P}_1$ and $\mathbf{P}_2$ associated with the Hamiltonian cycles H_1 and H_2, respectively, are

$$\mathbf{P}_1 = \begin{bmatrix} \cdot & 1 & \cdot & \cdot & \cdot & \cdot \\ \cdot & \cdot & \cdot & \cdot & \cdot & 1 \\ \cdot & \cdot & \cdot & 1 & \cdot & \cdot \\ \cdot & \cdot & \cdot & \cdot & 1 & \cdot \\ 1 & \cdot & \cdot & \cdot & \cdot & \cdot \\ \cdot & \cdot & 1 & \cdot & \cdot & \cdot \end{bmatrix} \quad \text{and} \quad \mathbf{P}_2 = \begin{bmatrix} \cdot & 1 & \cdot & \cdot & \cdot & \cdot \\ \cdot & \cdot & 1 & \cdot & \cdot & \cdot \\ \cdot & \cdot & \cdot & \cdot & \cdot & 1 \\ 1 & \cdot & \cdot & \cdot & \cdot & \cdot \\ \cdot & \cdot & \cdot & 1 & \cdot & \cdot \\ \cdot & \cdot & \cdot & \cdot & 1 & \cdot \end{bmatrix}.$$

Simple calculations give us

$$\det(\mathbf{I} - \mathbf{P}_1 + 1/6\mathbf{J}) = \det(\mathbf{I} - \mathbf{P}_2 + 1/6\mathbf{J}) = 6.$$

Let T_1 and T_2 be two subgraphs of the envelope graph, depicted respectively in Figures 2.14 and 2.15. Subgraph T_1 has one cycle, which is of length 3. Subgraph T_2 has one cycle, which is of length 4.

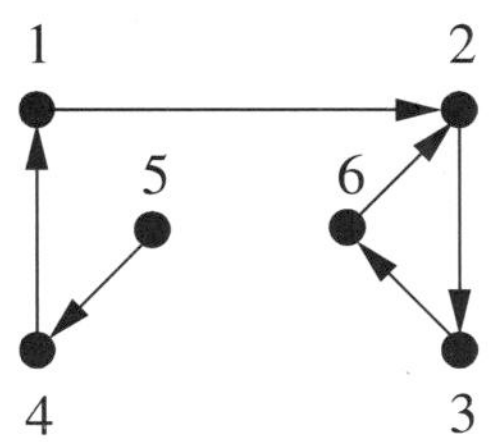

Fig. 2.14: Subgraph T_1 with one cycle

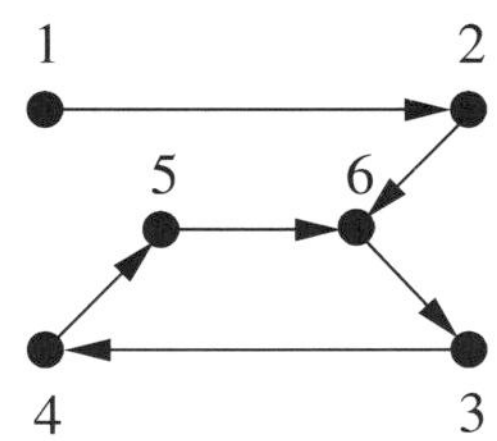

Fig. 2.15: Subgraph T_2 with one cycle

Because of their appearance, later on, we refer to subgraphs such as T_1 and T_2 as noose cycles. *The transition matrices $\mathbf{P}_3$ and $\mathbf{P}_4$ corresponding to subgraphs T_1 and T_2, respectively, are*

$$\mathbf{P}_3 = \begin{bmatrix} \cdot & 1 & \cdot & \cdot & \cdot & \cdot \\ \cdot & \cdot & 1 & \cdot & \cdot & \cdot \\ \cdot & \cdot & \cdot & \cdot & \cdot & 1 \\ 1 & \cdot & \cdot & \cdot & \cdot & \cdot \\ \cdot & \cdot & \cdot & 1 & \cdot & \cdot \\ \cdot & 1 & \cdot & \cdot & \cdot & \cdot \end{bmatrix} \quad \textit{and} \quad \mathbf{P}_4 = \begin{bmatrix} \cdot & 1 & \cdot & \cdot & \cdot & \cdot \\ \cdot & \cdot & \cdot & \cdot & \cdot & 1 \\ \cdot & \cdot & \cdot & 1 & \cdot & \cdot \\ \cdot & \cdot & \cdot & \cdot & 1 & \cdot \\ \cdot & \cdot & \cdot & \cdot & \cdot & 1 \\ \cdot & \cdot & 1 & \cdot & \cdot & \cdot \end{bmatrix}.$$

It is easy to check that

$$\det(\mathbf{I} - \mathbf{P}_3 + 1/6\mathbf{J}) = 3,$$

and

$$\det(\mathbf{I} - \mathbf{P}_4 + 1/6\mathbf{J}) = 4.$$

Let S_1 and S_2 be two subgraphs of the envelope, depicted in Figures 5.7 and 5.8. Each of subgraphs S_1 and S_2 has two disjoint cycles, both disjoint cycles in S_1 are of length 3, while one cycle in S_2 is of length 4 and the other is of length 2.

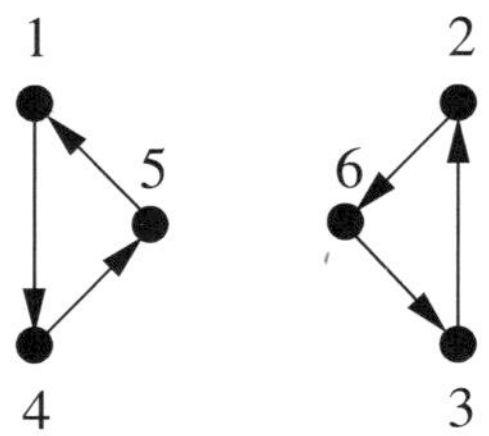

Fig. 2.16: Subgraph S_1

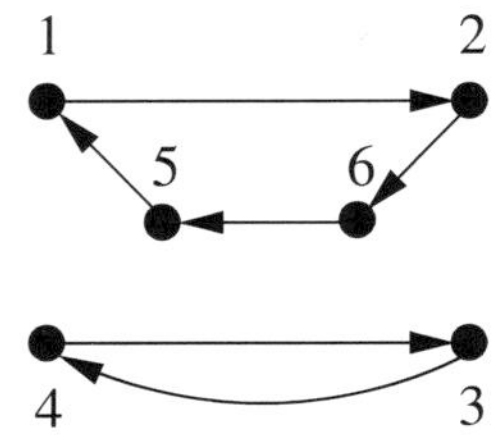

Fig. 2.17: Subgraph S_2

The transition matrices $\mathbf{P}_5$ *and* $\mathbf{P}_6$ *corresponding to subgraphs* T_1 *and* T_2*, respectively, are*

$$\mathbf{P}_5 = \begin{bmatrix} \cdot & \cdot & \cdot & 1 & \cdot & \cdot \\ \cdot & \cdot & \cdot & \cdot & \cdot & 1 \\ \cdot & 1 & \cdot & \cdot & \cdot & \cdot \\ \cdot & \cdot & \cdot & \cdot & 1 & \cdot \\ 1 & \cdot & \cdot & \cdot & \cdot & \cdot \\ \cdot & \cdot & 1 & \cdot & \cdot & \cdot \end{bmatrix} \quad \textit{and} \quad \mathbf{P}_6 = \begin{bmatrix} \cdot & 1 & \cdot & \cdot & \cdot & \cdot \\ \cdot & \cdot & \cdot & \cdot & \cdot & 1 \\ \cdot & \cdot & \cdot & 1 & \cdot & \cdot \\ \cdot & \cdot & 1 & \cdot & \cdot & \cdot \\ 1 & \cdot & \cdot & \cdot & \cdot & \cdot \\ \cdot & \cdot & \cdot & \cdot & 1 & \cdot \end{bmatrix}. \tag{2.9}$$

Again, it is easy to check that

$$\det(\mathbf{I} - \mathbf{P}_{S_1} + 1/6\mathbf{J}) = \det(\mathbf{I} - \mathbf{P}_{S_2} + 1/6\mathbf{J}) = 0. \tag{2.10}$$

This example suggests that Hamiltonian cycles, indeed, yield the maximal values of the determinant functional $\det \mathbf{W}(\mathbf{P})$.

In Example 2.3, it is not coincidental that the determinant functional has a value of 3 (or 4) at a subgraph containing a single cycle of length 3 (or 4), or that the determinant functional is zero at subgraphs containing two disjoint cycles. In Chapter 5, we prove that the determinant of $\mathbf{W}(\mathbf{P})$ is always k for a deterministic matrix $\mathbf{P}$ that corresponds to a subgraph containing a single cycle of length k. On the other hand, this determinant functional is always zero for a deterministic matrix $\mathbf{P}$ that corresponds to a subgraph containing two or more disjoint cycles.

In Chapter 3, we discuss the probabilistic interpretations of matrices $\mathbf{P}$, $\mathbf{W}(\mathbf{P})$, the inverse of the latter and other relevant matrices. These probabilistic interpretations exhibit the connection between the theory of Markov chains and the Hamiltonian cycle problem. Chapter 4 extends probabilistic approaches to the Hamiltonian cycle problem to include Markov decision processes. In Chapter 6, we use probabilistic arguments to prove that Hamiltonian cycles are minimisers of the trace functional of the inverse of a matrix similar to $\mathbf{W}(\mathbf{P})$.

Part II
Probabilistic Approaches

Chapter 3
Markov Chains

3.1 Introduction

Probabilistic methods have long been applied to solve discrete mathematics problems (see, for example, Erdős [39]–[40], and Alon and Spencer [3] for a recent and comprehensive treatment on probabilistic methods). Similarly, connections between Markov chains and graph theory have long been made (see Harary [59]). Our contribution here is to apply properties of Markov chains to the Hamiltonian cycle problem and to take advantage of the still emerging theory of perturbed Markov chains in this context. In Section 3.2, we give a brief introduction to Markov chains and various perturbations that we employ to obtain our results in this and subsequent chapters. More specifically, in Section 3.3, we present results on how fundamental matrices of Markov chains can be used to solve the Hamiltonian cycle problem, using their top-left matrix elements. In Section 3.4, we show that Hamiltonian cycles can be seen as variance minimisers of first hitting times and demonstrate a Hamiltonian gap that differentiates between the Hamiltonian and non-Hamiltonian graphs. Sections 3.3 and 3.4 make greater use of probabilistic methods than the remainder of this book. A reader unfamiliar with the latter could proceed to Chapter 4 and beyond with only a minimal loss of continuity.

3.2 Markov Chains and Perturbations

A *stochastic process* is a collection of random variables X_t, $t \geq 0$, which take values from a set $\mathcal{S}$ called a *state space*. There are two types of stochastic processes: *continuous-time* where $t \in [0, \infty)$, and *discrete-time* where $t \in \{0, 1, 2, \ldots\}$. In discrete-time case, each time point t is also called a *stage*. A stochastic process is *Markov* if the knowledge of the past does not influence the future, other than through the present. Mathematically, a stochastic process is Markov if, given $t_0 \leq t_1 \leq \cdots \leq t_n$, the following equality is satisfied for every set $\{X_{t_n} \leq x_n\}$

$$P\{X_{t_n} \leq x_n | X_{t_{n-1}}, \ldots, X_{t_1}\} = P\{X_{t_n} \leq x_n | X_{t_{n-1}}\}, \quad \text{for all } n \in \mathbb{N}.$$

Such a stochastic process is a *Markov chain* if $\mathcal{S}$ is discrete, and a *Markov process* otherwise. Significant applications of Markov chains and processes include those in geostatistics, stock market fluctuations, population processes, modelling games of chance such as Monopoly and, most recently, internet search engines including Google. Here, we are only concerned with finite discrete-time Markov chains, where $\mathcal{S} = \{1, 2, \ldots, N\}$ with $N = |\mathcal{S}| < \infty$ and $t \in \{0, 1, \ldots\}$. We recommend Kemeny and Snell [70] for an excellent introduction to finite Markov chains.

In other words, a discrete-time stochastic process X_t is a Markov chain if, given any $s_1, \ldots, s_n \in \mathcal{S}$ and $t_0 \leq t_1 \leq \cdots \leq t_n$, the following equality holds

$$P\{X_{t_n} = s_n | X_{t_{n-1}} = s_{n-1}, \ldots, X_{t_1} = s_1\} = P\{X_{t_n} = s_n | X_{t_{n-1}} = s_{n-1}\},$$

for all $n \in \mathbb{N}$. The *one-step transition probability*

$$P_{ij}^{t,t+1} = P\{X_{t+1} = j | X_t = i\} \tag{3.1}$$

is the probability that the system moves from state i at time t to state j at time $t+1$. If this probability does not depend on t, that is, if for $t \neq t'$,

$$P_{ij}^{t,t+1} = P_{ij}^{t',t'+1},$$

then the Markov chain X_t is *time-homogeneous*, with *stationary transition probabilities* P_{ij}. Unless otherwise indicated, all Markov chains we deal with in this book are time-homogeneous. The $N \times N$ *one-step probability transition matrix* $\mathbf{P}$ has entries $p_{ij} = P_{ij}$, where $p_{ij} \geq 0$ and $\sum_{j \in \mathcal{S}} p_{ij} = 1$. Similarly, for $i, j \in \mathcal{S}$ and for $t = 0, 1, 2, \ldots$, we define the *n-step transition probability*

$$P_{ij}^{(n)} = P\{X_{t+n} = j | X_t = i\}. \tag{3.2}$$

This conditional probability does not depend on t as we assume that all Markov chains considered are time-homogeneous. The $N \times N$ *n-step probability transition matrix* $\mathbf{P}^{(n)}$ has entries $p_{ij}^{(n)} = P_{ij}^{(n)}$. For $m, n \geq 0$, the matrix form of the Chapman-Kolmogorov equation is given by

$$\mathbf{P}^{(n+m)} = \mathbf{P}^{(m)}\mathbf{P}^{(n)}. \tag{3.3}$$

Following easily from the definitions above, (3.3) implies that the n-step probability transition matrix $\mathbf{P}^{(n)}$ can be obtained by multiplying the one-step transition matrix $\mathbf{P}$ by itself n times.

Every probability transition matrix is a *stochastic matrix*, that is, a matrix where the elements of each row are nonnegative and sum to 1. Consequently, every stochastic matrix is a probability transition matrix of some Markov chain. A *doubly stochastic matrix* is a stochastic matrix in which not only every row but also every column has a sum of unity. A *doubly stochastic deterministic matrix* is a *doubly stochastic matrix* every element of which is either 1 or 0.

Types of Markov Chains Consider a discrete state space $\mathcal{S}$, possibly countably infinite. If it is not possible to leave some state $i \in \mathcal{S}$, that is, $p_{ii} = 1$, then state i is *absorbing*. After starting at state i, if every return to i occurs in multiples of n_i steps, then i has a *period* n_i. Formally, we define $n_i = \gcd\{t : P\{X_t = i | X_0 = i\} > 0\}$, where gcd is the greatest common divisor. A state i is *aperiodic* if $n_i = 1$, and *periodic* with period n_i if $n_i > 1$.

If the probability of never returning to state i after starting at state i is positive, then the state i is *transient*. Formally, $P\{\mathcal{T}_i = \infty\} > 0$, where $\mathcal{T}_i = \inf\{t \geq 1 : X_t = i | X_0 = i\}$ is a random variable representing the first time of returning to i, and is also known as the *first return time* of state i. A state i that is not transient is said to be *recurrent*. For a recurrent state i, if the expectation of the returning time $\mathcal{T}_i$ is finite, that is, $\mathbb{E}[\mathcal{T}_i] < \infty$, then the state i is *positive recurrent*. Otherwise, it is *null recurrent*. A Markov chain is *absorbing* if it has at least one absorbing state.

A Markov chain is *irreducible* if there is a path with positive probability to go from any state to any other state. In an irreducible chain, if one state is recurrent (respectively, positive recurrent or null recurrent) then all states are likewise recurrent (respectively, positive recurrent or null recurrent). Finally, a Markov chain is *ergodic* (also known as *regular*) if it is irreducible and every state is positive recurrent. For finite $\mathcal{S}$, every recurrent state is positive recurrent. In particular, all states of a finite state irreducible Markov chain are positive recurrent.

Distribution Vectors A *distribution vector* $\boldsymbol{\nu}$ is a row vector with nonnegative entries ν_i, the probability of the system being in state i. An *initial distribution vector* $\boldsymbol{\nu}^{(0)}$ is a row vector with entries $\nu_i^{(0)}$, the probability of starting at state i. The distribution vector $\boldsymbol{\nu}^{(n)}$ with entries $\nu_i^{(n)}$, the probability of being at state i after n steps, is given by

$$\boldsymbol{\nu}^{(n)} = \boldsymbol{\nu}^{(0)}\mathbf{P}^n. \tag{3.4}$$

Each of the vectors $\boldsymbol{\nu}, \boldsymbol{\nu}^{(0)}$, and $\boldsymbol{\nu}^{(n)}$ sums to 1.

Stationary Distribution Matrix The *stationary distribution matrix* $\mathbf{P}^*$ is defined as follows

$$\mathbf{P}^* = \lim_{T \to \infty} \frac{1}{T+1} \sum_{t=0}^{T} \mathbf{P}^t. \tag{3.5}$$

It is well-known (see Doob [29]) that the limit in (3.5) exists and $\mathbf{P}^*$ is also known as the *Cesaro-limit matrix*, as it is the long-run average of the powers of $\mathbf{P}$. Every stationary distribution matrix $\mathbf{P}^*$ satisfies the following identity

$$\mathbf{P}^*\mathbf{P} = \mathbf{P}\mathbf{P}^* = \mathbf{P}^*\mathbf{P}^* = \mathbf{P}^*. \tag{3.6}$$

For *ergodic* (or *regular*) Markov chains, the following three properties hold.

(i) If the chain is aperiodic, the stationary distribution matrix $\mathbf{P}^*$ is equivalent to

$$\mathbf{P}^* = \lim_{t \to \infty} \mathbf{P}^t. \tag{3.7}$$

(ii) Every row of $\mathbf{P}^*$ is identical.
(iii) Define $\boldsymbol{e} = (1, \ldots, 1)^{\mathrm{T}} \in \mathbb{R}^N$ and let every row of $\mathbf{P}^*$ be the row-vector $\boldsymbol{q}$. Then, all entries of $\boldsymbol{q}$ are strictly positive and $\boldsymbol{q}$ is the unique solution to the system of linear equations

$$\begin{aligned} \boldsymbol{q}\mathbf{P} &= \boldsymbol{q} \\ \boldsymbol{q}\boldsymbol{e} &= 1. \end{aligned} \tag{3.8}$$

In this case, the vector $\boldsymbol{q}$ is often called the *stationary distribution vector* of the Markov chain.

Let $\mathbf{J} = \boldsymbol{e}\boldsymbol{e}^{\mathrm{T}}$, an $N \times N$ matrix of which every entry is unity. If $\mathbf{P}$ is doubly stochastic and induces an irreducible Markov chain, then $\boldsymbol{q} = 1/N\boldsymbol{e}^{\mathrm{T}}$ satisfies the system of equations (3.8), and consequently,

$$\mathbf{P}^* = 1/N\mathbf{J}. \tag{3.9}$$

In the context of this book, the Markov chains that we are interested in always correspond to an arbitrary but fixed graph G of order N. Such a graph defines a family of *induced Markov chains* that correspond to a family of $N \times N$ probability transition matrices $\mathbf{P}$ such that the entries of the ith row form a probability mass function on the edges of G emanating from vertex i of the graph, for $i = 1, 2, \ldots, N$.

Example 3.1 *We consider a 5-vertex graph depicted in Figure 3.1, and two Markov chains, each with the state space* $\mathcal{S} = \{1, 2, 3, 4, 5\}$.

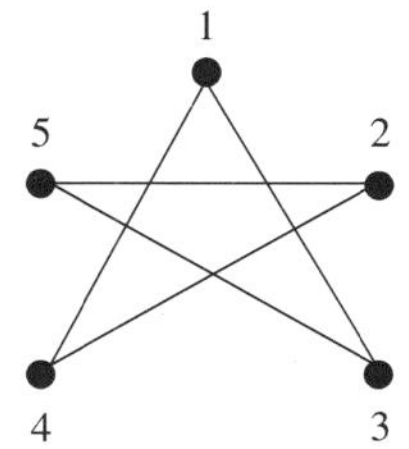

Fig. 3.1: A 5-vertex graph

In the first one, depicted in Figure 3.2, the system can travel from one state to exactly one other state. For example, from state 1, the system can go only to state 3. Therefore, the probability p_{13} *of going from state 1 to state 3 is 1, and* $p_{1i} = 0$ *for all* $i \neq 3, i \in \mathcal{S}$. *Of course, the corresponding graph displayed in Figure 3.2 is a spanning subgraph of the original graph from Figure 3.1.*

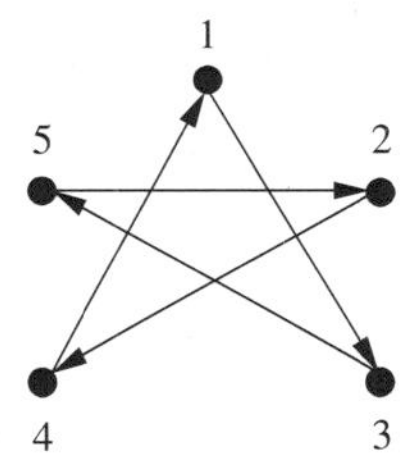

Fig. 3.2: An example of a finite discrete-time Markov chain with five states

The associated probability transition matrix $\mathbf{P}_1$ *is given by*

$$\mathbf{P}_1 = \begin{bmatrix} \cdot & \cdot & 1 & \cdot & \cdot \\ \cdot & \cdot & \cdot & 1 & \cdot \\ \cdot & \cdot & \cdot & \cdot & 1 \\ 1 & \cdot & \cdot & \cdot & \cdot \\ \cdot & 1 & \cdot & \cdot & \cdot \end{bmatrix}. \tag{3.10}$$

In this Markov chain, it is possible to travel from every state to any other state, hence it is irreducible. However, since every state has a period 5, the Markov chain is not aperiodic. Let $\boldsymbol{v} = (1/4, 1/4, 0, 1/4, 1/4)$ *be the initial distribution vector, then the distribution vector after one step is*

$$\boldsymbol{v}^{(1)} = \boldsymbol{v}\mathbf{P}_1 = [\,1/4 \; 1/4 \; 1/4 \; 1/4 \; 0\,].$$

After one step, the probabilities of the system being at states 1, 2, 3, 4 and 5 are $1/4, 1/4, 1/4, 1/4$ *and* 0, *respectively. The stationary distribution matrix* $\mathbf{P}_1^*$ *associated with this Markov chain is* $1/5\mathbf{J}$.

In the second Markov chain, depicted in Figure 3.3, from state 1, the system can travel to states 3 and 4, each with probability 1/2, and from state 5, the system can travel to state 2 with probability 1/5 and to state 3 with probability 4/5. From any other state, the system can only travel to exactly one other state.

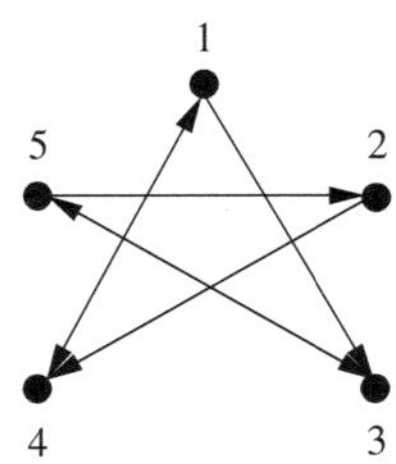

Fig. 3.3: Another example of a finite discrete-time Markov chain with five states

The associated probability transition matrix $\mathbf{P}_2$ *is given by*

$$\mathbf{P}_2 = \begin{bmatrix} \cdot & \cdot & \frac{1}{2} & \frac{1}{2} & \cdot \\ \cdot & \cdot & \cdot & 1 & \cdot \\ \cdot & \cdot & \cdot & \cdot & 1 \\ 1 & \cdot & \cdot & \cdot & \cdot \\ \cdot & \frac{1}{5} & \frac{4}{5} & \cdot & \cdot \end{bmatrix}. \tag{3.11}$$

This Markov chain is irreducible and aperiodic. With the initial distribution vector $\boldsymbol{v} = (1/4, 1/4, 0, 1/4, 1/4)$, *the distribution vector after one step is*

$$\boldsymbol{v}^{(1)} = \boldsymbol{v}\mathbf{P}_2 = [\,1/4\ 1/20\ 13/40\ 3/8\ 0\,].$$

The stationary distribution matrix $\mathbf{P}_2^*$ *associated with this Markov chain is* $\mathbf{P}_2^* = \boldsymbol{e}(2/15, 1/15, 1/3, 2/15, 1/3)$.

For every Markov chain, the matrix $\mathbf{I} - \mathbf{P} + \mathbf{P}^*$ is always invertible, and its inverse is called the *fundamental matrix*. Let $\mathbf{G}(\mathbf{P})$ be the fundamental matrix of a Markov chain specified by the probability transition matrix $\mathbf{P}$, then

$$\mathbf{G}(\mathbf{P}) = (\mathbf{I} - \mathbf{P} + \mathbf{P}^*)^{-1}. \tag{3.12}$$

Perturbations of Markov Chains We introduce two perturbations that have been applied to Markov chains in our line of research:

1. Symmetric linear perturbation ([34; 38; 37]): For a perturbation parameter $\varepsilon \in [0,1)$ and an $N \times N$ probability transition matrix $\mathbf{P}$, the perturbed matrix $\mathbf{P}^\varepsilon$ is defined as

$$\mathbf{P}^\varepsilon = (1-\varepsilon)\mathbf{P} + \varepsilon/N\mathbf{J}. \tag{3.13}$$

This symmetric linear perturbation ensures that the Markov chain specified by $\mathbf{P}^\varepsilon$ is always ergodic, while preserving double stochasticity whenever $\mathbf{P}$ is doubly stochastic.

2. Asymmetric linear perturbation ([49; 51; 35]): For a perturbation parameter $\varepsilon \in [0,1)$ and an $N \times N$ probability transition matrix $\mathbf{P}$, the perturbed matrix $\mathbf{P}^\varepsilon$ is defined as

$$\mathbf{P}^\varepsilon = (1-\varepsilon)\mathbf{P} + \varepsilon \begin{bmatrix} 1\,0 \cdots 0 \\ 0\,0 \cdots 0 \\ \vdots\;\vdots \quad\; \vdots \\ 0\,0 \cdots 0 \end{bmatrix} \mathbf{P} + \varepsilon \begin{bmatrix} 0\,0 \cdots 0 \\ 1\,0 \cdots 0 \\ \vdots\;\vdots\;\ddots\;\vdots \\ 1\,0 \cdots 0 \end{bmatrix}. \tag{3.14}$$

This asymmetric linear perturbation not only eliminates multiple ergodic classes but also differentiates vertex 1—referred to as the *home vertex* from Chapter 4 onwards—from other vertices. Additionally, it maintains roughly the level of sparsity of the original probability transition matrix $\mathbf{P}$.

Example 3.2 *We revisit a Markov chain introduced in Example 3.1, with the probability transition matrix* $\mathbf{P}_1$ *specified in (3.11)*

$$\mathbf{P}_1 = \begin{bmatrix} \cdot & \cdot & 1 & \cdot & \cdot \\ \cdot & \cdot & \cdot & 1 & \cdot \\ \cdot & \cdot & \cdot & \cdot & 1 \\ 1 & \cdot & \cdot & \cdot & \cdot \\ \cdot & 1 & \cdot & \cdot & \cdot \end{bmatrix}.$$

For $\varepsilon \in (0,1)$, *applying the symmetric linear perturbation defined in (3.13), we obtain*

$$\mathbf{P}^\varepsilon = \begin{bmatrix} \rho & \rho & 1-4/5\varepsilon & \rho & \rho \\ \rho & \rho & \rho & 1-4/5\varepsilon & \rho \\ \rho & \rho & \rho & \rho & 1-4/5\varepsilon \\ 1-4/5\varepsilon & \rho & \rho & \rho & \rho \\ \rho & 1-4/5\varepsilon & \rho & \rho & \rho \end{bmatrix},$$

where $\rho = 1/5\varepsilon$. *Applying the asymmetric linear perturbation defined in (3.14) gives us*

$$\mathbf{P}^{\varepsilon} = \begin{bmatrix} \cdot & \cdot & 1 & \cdot & \cdot \\ \varepsilon & \cdot & \cdot & 1-\varepsilon & \cdot \\ \varepsilon & \cdot & \cdot & \cdot & 1-\varepsilon \\ 1 & \cdot & \cdot & \cdot & \cdot \\ \varepsilon & 1-\varepsilon & \cdot & \cdot & \cdot \end{bmatrix}.$$

3.3 Hitting Times and the Fundamental Matrix

In this section, we derive important relationships between entries of the fundamental matrix $\mathbf{G}(\mathbf{P})$ (associated with the Markov chain defined by the transition matrix $\mathbf{P}$) and the moments of the first hitting times, with respect to a given initial distribution. Indeed, we focus on τ_1, the random variable denoting the first return time of the home state/vertex 1, given that Markov chain starts at that vertex, and we denote by $Var[\tau_1]$ its variance. We show that whenever $\mathbf{P}$ is doubly stochastic the first diagonal entry of $\mathbf{G}(\mathbf{P})$ has the linear form $a + bVar[\tau_1]$, where a and b depend only on N. Of course, if we take vertex k as the home vertex, then the kth diagonal entry of $\mathbf{G}(\mathbf{P})$ can be shown to have an analogous linear form $a + bVar[\tau_k]$.

The above is significant because, subsequently, we shall show that the Hamiltonian cycle problem is equivalent to an optimisation problem, the objective of which is to minimise the top-left element of the fundamental matrix of the Markov chain permissible on a given graph, over the space of Markov chains on the graph that have doubly stochastic transition matrices. Therefore, the HCP is also equivalent to the problem of minimising the variance of τ_1 over the space of doubly stochastic probability transition matrices $\mathbf{P}$ that can be associated with the given graph.

We prove this equivalence by deriving the formulae for the entries of the first column of the fundamental matrix. Consider $\boldsymbol{w} = (\mathbf{I} - \mathbf{P} + \mathbf{P}^*)^{-1}\boldsymbol{r}$, the first column of $\mathbf{G}(\mathbf{P})$, where $\boldsymbol{r} = (1, 0, \ldots, 0)^{\mathrm{T}}$ is an N-dimensional column vector. Then, $(\mathbf{I} - \mathbf{P} + \mathbf{P}^*)\boldsymbol{w} = \boldsymbol{r}$. Recall that for irreducible Markov chains, every row of $\mathbf{P}^*$ is identical to $\boldsymbol{q}$, the stationary distribution vector of $\mathbf{P}$. Therefore, $\mathbf{P}^*\boldsymbol{w} = (\sum_i q_i w_i)\,\boldsymbol{e}$ and hence

$$\boldsymbol{w} - \mathbf{P}\boldsymbol{w} + (\sum_i q_i w_i)\boldsymbol{e} = \boldsymbol{r}. \tag{3.15}$$

We denote by $\mathbb{E}_i[\tau_1]$ the expectation of the first return time to vertex 1, starting from vertex i, for $i \in \mathcal{S}$. Before stating our results, we briefly introduce some concepts and recall some well-known theorems.

A *σ-field* $\mathcal{F}$ over a set Ω is a family of subsets of Ω such that it satisfies the three conditions:

(i) $\Omega \in \mathcal{F}$,
(ii) if $Y \in \mathcal{F}$ then $Y^c \in \mathcal{F}$, where Y^c is the complement of Y,
(iii) a countable union of sets in $\mathcal{F}$ is also in $\mathcal{F}$.

A real random variable X is a measurable function on $(\Omega, \mathcal{F})$, that is, for all intervals $A \subset \mathbb{R}$, the set $\{\omega \in \Omega : X(\omega) \in A\} \in \mathcal{F}$. Let $\mathcal{F}_1 \subset \mathcal{F}_2 \subset \mathcal{F}_3 \subset \cdots$ be an increasing family of σ-fields. Then, a stochastic process $\{X_n\}$ taking values in $\mathbb{R}$ is said to be a *martingale* with respect to $\{\mathcal{F}_n\}$ if

(i) X_n is $\mathcal{F}_n$-measurable for all n, and
(ii) $\mathbb{E}[|X_n|] < \infty$ and $\mathbb{E}[X_{n+1}|\mathcal{F}_n] = X_n$ for all n.

The family $\{\mathcal{F}_n\}$ is often understood from the context, $\mathcal{F}_n = \sigma(X_1, \ldots, X_n)$ being a common choice. A random variable τ taking values in $\{0, 1, 2, \ldots, \infty\}$ is a *stopping time* with respect to $\{\mathcal{F}_n\}$ if $\{\tau \leq n\} \in \mathcal{F}_n$ for all n. Equivalently, $\{\tau = n\} \in \mathcal{F}_n$ for all n. Intuitively, at each n, based on the "observed history" at each n, one knows whether τ has occurred or not.

Let $X_0 = i$ and $p(X_{m-1}, j)$ be the (X_{m-1}, j)th element of the matrix $\mathbf{P}$. We define

$$M_n = \sum_{m=1}^{n} \{w_{X_m} - \sum_j p(X_{m-1}, j) w_j\}, \text{ for } n \geq 1,$$

where w_{X_m} is the X_mth entry of the vector $\boldsymbol{w}$. Then, it is well-known (see Borkar [16, Chapter 3]) that the sequence $\{M_n\}$ is a martingale with respect to the family of σ-fields $\mathcal{F}_n = \sigma(X_i, i \leq n)$, that is, the σ-field generated by the sets of the type $\{\omega : X_i \in A\}$ for $i \leq n$ and intervals $A \subset \mathbb{R}$. The following result, by Doob, can be found in Borkar [16, Chapter 3].

Theorem 3.1. ***Optional Sampling theorem.*** *Let X_i be a martingale with respect to $\{\mathcal{F}_i\}$ and η a bounded stopping time with respect to $\{\mathcal{F}_i\}$. Then*

$$\mathbb{E}[X_\eta] = \mathbb{E}[X_1].$$

Now we are ready to state results concerning the elements of $\boldsymbol{w}$, the first column of the fundamental matrix $\mathbf{G}(\mathbf{P})$.

Theorem 3.2. *The entries w_i of the vector $\boldsymbol{w}$ are given by*

$$w_1 = \frac{1}{2} \frac{\mathbb{E}_1[\tau_1(\tau_1 + 1)]}{\mathbb{E}_1[\tau_1]^2} \tag{3.16}$$

$$w_j = \frac{1}{2} \frac{\mathbb{E}_1[\tau_1(\tau_1 + 1)]}{\mathbb{E}_1[\tau_1]^2} - \frac{\mathbb{E}_j[\tau_1]}{\mathbb{E}_1[\tau_1]}, \quad \textit{for } j \neq 1. \tag{3.17}$$

Proof. By the Optional Sampling theorem, for $i \in \mathcal{S}$ and $n \geq 1$,

$$\mathbb{E}_i[M_{\tau_1 \wedge n}] = \mathbb{E}_i[\sum_{m=1}^{\tau_1 \wedge n} (w_{X_m} - \sum_j p(X_{m-1}, j) w_j)] = 0,$$

where $\tau_1 \wedge n = \min\{\tau_1, n\}$. Since $0 < \mathbb{E}_i[\tau_1] < \infty$, we take the limit as n tends to ∞ and apply the dominated convergence theorem to obtain

$$\mathbb{E}_i[\sum_{m=1}^{\tau_1} (w_{X_m} - \sum_j p(X_{m-1}, j) w_j)] = 0. \tag{3.18}$$

Recall that $X_0 = i$, and $X_{\tau_1} = 1$. Then, for each $i = 1, 2, \ldots, n$, the left-hand side of (3.18) becomes

$$\begin{aligned}
&\mathbb{E}_i[\sum_{m=0}^{\tau_1 - 1} (w_{X_{m+1}} - \sum_j p(X_m, j) w_j)] \\
&= \mathbb{E}_i[\sum_{m=0}^{\tau_1 - 1} (w_{X_m} - \sum_j p(X_m, j) w_j) + w_{X_{\tau_1}} - w_{X_0}] \\
&= w_1 + \mathbb{E}_i[\sum_{m=0}^{\tau_1 - 1} (w_{X_m} - \sum_j p(X_m, j) w_j)] - w_i \\
&= w_1 - \mathbb{E}_i[\sum_{m=0}^{\tau_1 - 1} (\sum_j p(X_m, j) w_j - w_{X_m})] - w_i.
\end{aligned}$$

Since the right-hand side of (3.18) is 0,

$$w_1 - \mathbb{E}_i[\sum_{m=0}^{\tau_1 - 1} (\sum_j p(X_m, j) w_j - w_{X_m})] - w_i = 0. \tag{3.19}$$

Left-multiplying both sides of (3.15) by the limiting matrix $\mathbf{P}^*$ yields

$$\mathbf{P}^* \{\boldsymbol{w} - \mathbf{P}\boldsymbol{w} + (\sum_i q_i w_i)\boldsymbol{e}\} = \mathbf{P}^* \boldsymbol{r}. \tag{3.20}$$

As $\mathbf{P}^*\mathbf{P} = \mathbf{P}^*$, the left-hand side of (3.20) becomes

$$\begin{aligned}
&\mathbf{P}^*\boldsymbol{w} - \mathbf{P}^*\mathbf{P}\boldsymbol{w} + \mathbf{P}^*(\sum_i q_i w_i)\boldsymbol{e} \\
&= \mathbf{P}^*\boldsymbol{w} - \mathbf{P}^*\boldsymbol{w} + \mathbf{P}^*(\sum_i q_i w_i)\boldsymbol{e} \\
&= \mathbf{P}^*(\sum_i q_i w_i)\boldsymbol{e}.
\end{aligned}$$

Therefore, by (3.20), $\mathbf{P}^*(\sum_i q_i w_i)\boldsymbol{e} = \mathbf{P}^*\boldsymbol{r}$, and, consequently, $\sum_i q_i w_i = q_1$. Using this fact, the X_mth row of (3.15) is

$$w_{X_m} - \sum_j p(X_m, j)w_j + q_1 = I\{X_m = 1\},$$

where $I\{X_m = 1\}$ is 1 if $X_m = 1$ and 0 otherwise, as $r_{X_m} = 1$ if $X_m = 1$ and 0 otherwise. Therefore,

$$w_{X_m} - \sum_j p(X_m, j)w_j = I\{X_m = 1\} - q_1,$$

and (3.19) becomes

$$w_1 - \mathbb{E}_i[\sum_{m=0}^{\tau_1 - 1} (w_{X_m} - I\{X_m = 1\} + q_1 - w_{X_m})] - w_i = 0$$
$$w_1 + \mathbb{E}_i[\sum_{m=0}^{\tau_1 - 1} (I\{X_m = 1\} - q_1)] - w_i = 0.$$

Rearranging the last equation yields

$$w_i = w_1 + \mathbb{E}_i[\sum_{m=0}^{\tau_1 - 1} (I\{X_m = 1\} - q_1)]. \tag{3.21}$$

For all $i \neq 1, i \in \mathcal{S}$ and $m < \tau_1$, $X_m \neq 1$ as τ_1 is the first hitting time of vertex 1. Therefore, for $i \neq 1, i \in \mathcal{S}$, (3.21) is the same as

$$\begin{aligned} w_i &= w_1 + \mathbb{E}_i[\sum_{m=0}^{\tau_1 - 1} (-q_1)] \\ &= w_1 + \mathbb{E}_i[-q_1\tau_1] \\ &= w_1 - q_1\mathbb{E}_i[\tau_1]. \end{aligned} \tag{3.22}$$

For $i = 1$, $I\{X_0 = 1\} = 1$, and (3.21) reduces to $w_1 = w_1$ since

$$\begin{aligned} w_1 &= w_1 + \mathbb{E}_1[\sum_{m=0}^{\tau_1 - 1} (-q_1) + 1] \\ &= w_1 + \mathbb{E}_i[-q_1\tau_1] + 1 \\ &= w_1 - q_1\mathbb{E}_1[\tau_1] + 1 \\ &= w_1, \end{aligned} \tag{3.23}$$

the second last equality comes from the fact that $q_1\mathbb{E}_1[\tau_1] = 1$, by Borkar [16, Theorem 5.3.2]. As $\sum_i q_i w_i = q_1$, we multiply both sides of (3.22) by q_i, and sum over all $i \neq 1$ to obtain

$$\begin{aligned}
\sum_{i\neq 1} w_i q_i &= w_1 \sum_{i\neq 1} q_i - q_1 \sum_{i\neq 1} q_i \mathbb{E}_i[\tau_1] \\
\sum_{i} w_i q_i - w_1 q_1 &= w_1 \sum_{i\neq 1} q_i - q_1 \sum_{i\neq 1} q_i \mathbb{E}_i[\tau_1] \\
q_1 &= w_1 \sum_{i} q_i - q_1 \sum_{i\neq 1} q_i \mathbb{E}_i[\tau_1] \\
&= w_1 - q_1 \sum_{i\neq 1} q_i \mathbb{E}_i[\tau_1], \qquad (3.24)
\end{aligned}$$

with the last equality obtained by the property of the stationary distribution vector $\boldsymbol{q}$ (see (3.8)). As $q_1\mathbb{E}_1[\tau_1] = 1$, $q_1 = \mathbb{E}_1[\tau]^{-1}$, and (3.24) becomes

$$\begin{aligned}
q_1 &= w_1 - q_1 \sum_{i} q_i \mathbb{E}_i[\tau_1] + q_1 q_1 \mathbb{E}_1[\tau_1] \\
&= w_1 - q_1 \sum_{i} q_i \mathbb{E}_i[\tau_1] + q_1.
\end{aligned}$$

Consequently,

$$w_1 = q_1 \sum_{i} q_i \mathbb{E}_i[\tau_1]. \qquad (3.25)$$

By [16, Theorem 5.3.4], equation (3.25) is equivalent to

$$\begin{aligned}
w_1 &= q_1 \frac{\mathbb{E}_1[\sum_{m=0}^{\tau_1-1} \mathbb{E}_{X_m}[\tau_1]]}{\mathbb{E}_1[\tau_1]} \\
&= \mathbb{E}[\tau_1]^{-1} \frac{\mathbb{E}_1[\sum_{m=0}^{\tau_1-1} \mathbb{E}_{X_m}[\tau_1]]}{\mathbb{E}_1[\tau_1]} \\
&= \frac{\mathbb{E}_1[\sum_{m=0}^{\tau_1-1} \mathbb{E}_{X_m}[\tau_1]]}{\mathbb{E}_1[\tau_1]^2} \\
&= \frac{\mathbb{E}_1[\sum_{m=0}^{\tau_1-1} (\tau_1 - m)]}{\mathbb{E}_1[\tau_1]^2} \\
&= \frac{1}{2} \frac{\mathbb{E}_1[\tau_1(\tau_1+1)]}{\mathbb{E}_1[\tau_1]^2},
\end{aligned}$$

with the last equation obtained by using the occupational measure identity in Pitman [85, p.74]. Hence, we complete the proof for (3.16).

From (3.16) and (3.22), for $i \neq 1$, we obtain

$$w_i = w_1 - q_1 \mathbb{E}_i[\tau_1]$$

$$= \frac{1}{2}\frac{\mathbb{E}_1[\tau_1(\tau_1+1)]}{\mathbb{E}_1[\tau_1]^2} - \mathbb{E}_1[\tau_1]^{-1}\mathbb{E}_i[\tau_1]$$
$$= \frac{1}{2}\frac{\mathbb{E}_1[\tau_1(\tau_1+1)]}{\mathbb{E}_1[\tau_1]^2} - \frac{\mathbb{E}_i[\tau_1]}{\mathbb{E}_1[\tau_1]},$$

which completes the proof for (3.17). □

The following theorem gives analytic expressions for w_1, the top-left element of the fundamental matrix $\mathbf{G}(\mathbf{P})$ for doubly stochastic matrices $\mathbf{P}$. In effect, it shows that this element can be regarded as an objective function

$$w_1 = w_1(\mathbf{P}) = a + bVar[\tau_1|\mathbf{P}].$$

In the next section, the associated optimisation problem will be discussed in much more detail.

Theorem 3.3. *For a given doubly stochastic* $\mathbf{P}$, *we have*

$$w_1 = \frac{1}{N^2}\sum_i \mathbb{E}_i[\tau_1] = \frac{(N+1)}{2N} + \frac{1}{2N^2}\mathbb{E}_1[(\tau_1 - N)^2]. \tag{3.26}$$

Moreover, if $\mathbf{P}$ *is associated with a Hamiltonian cycle and we apply the symmetric linear perturbation defined in (3.13), then* w_1 *simplifies to*

$$w_1^{\varepsilon} = \frac{1}{2}\frac{(N+1)}{N} + O(\varepsilon). \tag{3.27}$$

Proof. Recall that for a doubly stochastic probability transition matrix $\mathbf{P}$, the stationary distribution vector $\boldsymbol{q}$ of $\mathbf{P}$ is $1/N\boldsymbol{e}^{\mathrm{T}}$, so $q_i = 1/N$ for all i. Since $q_1 = \mathbb{E}_1[\tau_1]^{-1}$, we have $\mathbb{E}_1[\tau_1] = N$. By (3.25),

$$w_1 = q_1\sum_i q_i\mathbb{E}_i[\tau_1] = 1/N^2\sum_i \mathbb{E}_i[\tau_1].$$

In addition, by (3.16),

$$w_1 = \frac{\mathbb{E}_1[\tau_1(\tau_1+1)]}{2(\mathbb{E}_1[\tau_1])^2}$$
$$= \frac{\mathbb{E}_1[\tau_1^2] + \mathbb{E}_1[\tau_1]}{2(\mathbb{E}_1[\tau_1])^2}$$
$$= \frac{\mathbb{E}_1[(\tau_1 - \mathbb{E}_1[\tau_1])^2] + \mathbb{E}_1[\tau_1]^2 + \mathbb{E}_1[\tau_1]}{2(\mathbb{E}_1[\tau_1])^2}$$
$$= \frac{\mathbb{E}_1[\tau_1]^2 + \mathbb{E}_1[\tau_1]}{2(\mathbb{E}_1[\tau_1])^2} + \frac{\mathbb{E}_1[(\tau_1 - \mathbb{E}_1[\tau_1])^2]}{2(\mathbb{E}_1[\tau_1])^2}$$
$$= \frac{N+1}{2N} + \frac{\mathbb{E}_1[(\tau_1 - N)^2]}{2N^2}.$$

For the second part of Theorem 3.3, we need to show that $\mathbb{E}[(\tau_1 - N)^2] = O(\varepsilon)$. The probability that the system travels using at least one weak transition, that is, an edge with probability ε, in N steps is at most $N(N-1)\varepsilon$. Therefore, $P\{\tau_1 \neq N\} \leq N(N-1)\varepsilon$. Also, for the inequality $\tau_1 > kN$ to hold for $k \geq 1$, the system must travel using at least one weak transition in each block of N consecutive steps. Consequently, $P\{\tau_1 > kN\} \leq [N(N-1)\varepsilon]^k$. This leads to

$$\begin{aligned}
\mathbb{E}[(\tau_1 - N)^2] &= \sum_{j \geq 1} j^2 P\{|\tau_1 - N| = j\} \\
&= \sum_{j \geq 1} j^2 P\{\tau_1 = N - j\} + \sum_{j \geq 1} j^2 P\{\tau_1 = N + j\} \\
&\leq N^2 \sum_{j \geq 1} P\{\tau_1 = N - j\} + \sum_{j \geq 1} j^2 P\{\tau_1 = N + j\} \\
&\leq N^2 P\{\tau_1 \neq N\} + \sum_{j \geq 1} j^2 P\{\tau_1 = N + j\} \\
&\leq N^3 (N-1)\varepsilon + \sum_{j \geq 1} j^2 P\{\tau_1 = j + N\} \\
&\leq N^3 (N-1)\varepsilon + \sum_{k \geq 1} [(k+1)N]^2 P\{\tau_1 > kN\} \\
&\leq N^3 (N-1)\varepsilon + \sum_{k \geq 1} [(k+1)N]^2 [N(N-1)\varepsilon]^k \\
&\leq O(\varepsilon).
\end{aligned}$$

This completes the proof. □

3.4 Hamiltonian Cycles as Hitting Time Variance Minimisers

Consider a given graph G with N vertices and recall that every unperturbed probability transition matrix $\mathbf{P}$ induced by G has the property that $p_{ij} = 0$ whenever edge (i, j) is not present in the graph. All unperturbed probability transition matrices considered in this section are assumed to be induced by G. It is clear that if $\mathbf{P}$ is the probability transition matrix induced by a Hamiltonian cycle, then starting from the home vertex 1, the corresponding Markov chain will return to it after exactly N transitions, implying that $Var[\tau_1|\mathbf{P}] = 0$. Furthermore, since in such a case the symmetric linear perturbation defined in (3.13) does not alter the ergodic structure of this chain, it is reasonable to expect that $Var[\tau_1|\mathbf{P}^{\varepsilon}]$ tends to 0 as ε approaches 0. Hence, in view of Theorem 3.3, it is also reasonable to conjecture that probability transition matrices $\mathbf{P}^{\varepsilon}_{\text{H}}$ induced by Hamiltonian cycles achieve the minimum in the optimisation problem

$$\min Var[\tau_1|\mathbf{P}^\varepsilon],$$

over the space of doubly stochastic probability transition matrices $\mathbf{P}^\varepsilon$ induced by the given graph G, provided that $\varepsilon > 0$ is sufficiently small. Equivalently, we conjecture that probability transition matrices $\mathbf{P}^\varepsilon_{\mathrm{H}}$ induced by Hamiltonian cycles achieve the minimum in the optimisation problem

$$\min w_1^\varepsilon(\mathbf{P}^\varepsilon),$$

over the same space of doubly stochastic probability transition matrices, provided that $\varepsilon > 0$ is sufficiently small. This section is devoted to proving these conjectures and to establishing the existence of the *Hamiltonicity gap* property, which demonstrates that the optimal values of these two objective functions can be used to distinguish between Hamiltonian and non-Hamiltonian graphs, without actually requiring that a Hamiltonian cycle be found.

We denote by $\mathbf{P}^\varepsilon$ the probability transition matrix obtained from $\mathbf{P}$ after we apply the symmetric linear perturbation defined in (3.13), by $\mathcal{D}_d$ the finite set of perturbed doubly stochastic deterministic matrices, and by $\mathcal{D}$ the convex set of perturbed doubly stochastic matrices obtained by taking the closed convex hull of $\mathcal{D}_d$. We also write $\mathcal{D}_d$ as the disjoint union $\mathcal{D}_{\mathrm{H}} \cup \mathcal{D}_s$, where $\mathcal{D}_{\mathrm{H}}$ represents the set of Hamiltonian cycles and $\mathcal{D}_s$ the set of disjoint unions of short cycles that cover the graph. All sets $\mathcal{D}_d, \mathcal{D}, \mathcal{D}_{\mathrm{H}}$ and $\mathcal{D}_s$ depend on the perturbation parameter ε, which we suppress. Likewise, we suppress the dependence of aforementioned objective functions on $\mathbf{P}^\varepsilon$, except where ambiguity might arise.

Lemma 3.1. *For $\mathbf{P}^\varepsilon \in \mathcal{D}_s$, the top-left entry w_1^ε of the fundamental matrix $\mathbf{G}(\mathbf{P}^\varepsilon)$ tends to infinity as ε approaches* 0.

Proof. Let $i \in \mathcal{S}$ lie in a short cycle of $\mathbf{P}$ not containing vertex 1. Then, a chain starting at i must make an $\varepsilon-$transition before ever hitting 1. Thus, if ζ denotes the first time it makes an $\varepsilon-$transition, then

$$\begin{aligned}\mathbb{E}_i[\tau_1] &\geq \mathbb{E}_i[\zeta] \\ &= \sum_{m\geq 1} m(N-1)\varepsilon\{1-(N-1)\varepsilon\}^{m-1} \\ &= 1/\{(N-1)\varepsilon\}.\end{aligned}$$

The claim follows from (3.26). □

We say that $\mathbf{P} \in \mathcal{D}$ is a perturbation of Hamiltonian cycle if there exists a $\hat{\mathbf{P}} \in \mathcal{D}_{\mathrm{H}}$ such that $||\mathbf{P} - \hat{\mathbf{P}}|| = C\varepsilon_0$ for prescribed $C, \varepsilon_0 > 0$. Let $\mathcal{D}_p$ denote the set of such matrices $\mathbf{P}$.

Theorem 3.4. *For sufficiently small $\varepsilon > 0$, all minima of w_1^ε on $\mathcal{D}$ are attained on $\mathcal{D}_p$.*

The proof uses the following lemma. We use the notation $\theta(1)$ to denote any function $g(\varepsilon)$ satisfying $\liminf_{\varepsilon\downarrow 0} g(\varepsilon) > 0$ and $\limsup_{\varepsilon\downarrow 0} g(\varepsilon) < \infty$.

Lemma 3.2. *For $\mathbf{P} \in \mathcal{D} \setminus \mathcal{D}_p$,*

$$w_1^\varepsilon \geq \frac{1}{2}\frac{(N+1)}{N} + \theta(1).$$

Proof. By (3.26)

$$w_1^\varepsilon = \frac{1}{2}\frac{N+1}{N} + \frac{1}{2N^2}\mathbb{E}_1[(\tau_1 - N)^2].$$

Thus, it suffices to prove that for $\mathbf{P} \in \mathcal{D} \setminus \mathcal{D}_p$,

$$\mathbb{E}_1[(\tau_1 - N)^2] = \theta(1),$$

which it indeed will be if the weighted digraph of $\mathbf{P}$ contains a short (that is, non-Hamiltonian) cycle containing 1 which has a probability $\theta(1)$. Now, $\mathbf{P}$ is a finite convex combination of elements of $\mathcal{D}_e$, with either

(i) weight $\theta(1)$ for at least one $\hat{\mathbf{P}} \in \mathcal{D}_s$, or
(ii) weights $\theta(1)$ for at least two distinct $\mathbf{P}', \mathbf{P}'' \in \mathcal{D}_{\mathrm{H}}$.

It is easy to see that if (i) and (ii) were false, $\mathbf{P}$ would be in $\mathcal{D}_p$. In case of (i), it is clear that $\mathbf{P}$ has a short cycle containing 1 which has a probability $\theta(1)$. In case of (ii), let $\mathbf{P}$ put the $\theta(1)$ weights on $\mathbf{P}_1, \mathbf{P}_2 \in \mathcal{D}_{\mathrm{H}}$, corresponding to two Hamiltonian cycles $1 \to x_1 \to x_2 \to \cdots \to x_{N-1} \to 1$ and $1 \to y_1 \to y_2 \to \cdots \to y_{N-1} \to 1$ respectively. Let $i = \min\{j > 1 : x_j \neq y_j\}$. Then, $x_i = y_k$ for some $k > i$. Hence, $1 \to \cdots \to x_i \to y_{k+1} \to \cdots \to y_{s-1} \to 1$ defines a short cycle that contains 1 and has a probability $\theta(1)$. This completes the proof. □

Proof of Theorem 3.4. Since w^ε is the unique solution to the well-posed linear system given by (3.15), it depends smoothly on the coefficients thereof by Cramer's rule, hence on ε. From Theorem 3.3, it then follows that

$$w_1^\varepsilon = (N+1)/(2N) + \mathcal{O}(\varepsilon)$$

for $\mathbf{P} \in \mathcal{D}_p$. The claim now follows from Lemma 3.2. □

Directional derivatives Before showing that the minima of w_1^ε on $\mathcal{D}$ are attained at doubly stochastic matrices induced by Hamiltonian cycles, we need to present some results on directional derivatives. We now derive an expression for the directional derivative of our objective function w_1^ε.

Let $\mathbf{P}_0$ and $\mathbf{P}_1$ denote two doubly stochastic matrices in $\mathcal{D}$ and for $0 \leq \lambda \leq 1$, define $\mathbf{P}_\lambda = \lambda\mathbf{P}_1 + (1-\lambda)\mathbf{P}_0$. Correspondingly, define $\kappa_\lambda(i) = \mathbb{E}_i[\tau_1]$ for $i \in \mathcal{S}$, where the dependence of the distribution of τ_1 on the parameter λ is implicit. Also, let $\tilde{\mathbf{P}}_0, \tilde{\mathbf{P}}_1$ and $\tilde{\mathbf{P}}_\lambda$ denote sub-matrices derived, respectively, from $\mathbf{P}_0, \mathbf{P}_1$, and $\mathbf{P}_\lambda$ by deletion of their first row and column. Similarly, for vectors we use tilde to denote truncations resulting from the omission of the first entry, for instance, $\tilde{\boldsymbol{\kappa}}_\lambda = (\kappa_\lambda(2), \ldots, \kappa_\lambda(N))^{\mathrm{T}}$, $\tilde{\mathbf{e}} = (1, \ldots, 1)^{\mathrm{T}} \in \mathbb{R}^{N-1}$.

By Borkar [15, Lemma 1.3], we know that $\tilde{\boldsymbol{\kappa}}_\lambda$ is the unique solution to

$$\tilde{\boldsymbol{\kappa}}_\lambda = \tilde{\mathbf{e}} + \tilde{\mathbf{P}}_\lambda \tilde{\boldsymbol{\kappa}}_\lambda, \tag{3.28}$$

that is,

$$\tilde{\boldsymbol{\kappa}}_\lambda = (\mathbf{I} - \tilde{\mathbf{P}}_\lambda)^{-1}\tilde{\mathbf{e}}. \tag{3.29}$$

We denote by $\{X_n\}$ the Markov chain governed by $\mathbf{P}_\lambda$, and by p_{ij} the elements of $\mathbf{P}_\lambda$.

Lemma 3.3. *Let $\nu_\lambda(i) = \mathbb{E}[\sum_{m=1}^{\tau_1} I\{X_m = i\}]$ when the initial distribution is the uniform distribution. Then, $\tilde{\boldsymbol{\nu}}_\lambda^{\mathrm{T}} = (\nu_\lambda(2), \ldots, \nu_\lambda(N))$ is the unique solution to*

$$\tilde{\boldsymbol{\nu}}_\lambda^{\mathrm{T}} = 1/N\tilde{\mathbf{e}}^{\mathrm{T}} + \tilde{\boldsymbol{\nu}}_\lambda^{\mathrm{T}}\tilde{\mathbf{P}}_\lambda = 1/N\tilde{\mathbf{e}}^{\mathrm{T}}(\mathbf{I} - \tilde{\mathbf{P}}_\lambda)^{-1}. \tag{3.30}$$

Proof. We define $\zeta_1 = \min\{n \geq 0 : X_n = 1\}$ ($= \infty$ if this set is empty), and $\nu_\lambda^j(i) = \mathbb{E}_j[\sum_{m=0}^{\zeta_1} I\{X_m = i\}]$. Then, for $j \neq 1$, $\zeta_1 = \tau_1$, and

$$\nu_\lambda^j(i) = \mathbb{E}_j[\sum_{m=0}^{\zeta_1} I\{X_m = i\}] = \mathbb{E}_j[\sum_{m=0}^{\tau_1} I\{X_m = i\}].$$

Consider $i, j \neq 1$. We note that for $n \geq 1$, $M_n = \sum_{m=1}^{n}(I\{X_m = i\} - p_{X_{m-1},i})$ is a martingale. By the Optional Sampling theorem (stated in Theorem 3.1) and for $T \geq 1$,

$$\mathbb{E}_j[\sum_{m=1}^{\tau_1 \wedge T}(I\{X_m = i\} - p_{X_{m-1},i})] = 0 \tag{3.31}$$

Letting $T \uparrow \infty$ and using the easily established fact that $\mathbb{E}[\tau_1^2] < \infty$, which implies uniform integrability of the sum above as T varies, we have

$$\mathbb{E}_j[\sum_{m=1}^{\tau_1}(I\{X_m = i\} - p_{X_{m-1},i})] = 0.$$

Thus,

$$\begin{aligned}
\nu_\lambda^j(i) &= \mathbb{E}_j[\sum_{m=0}^{\tau_1} I\{X_m = i\}] \\
&= \delta_{ij} + \mathbb{E}_j[\sum_{m=1}^{\tau_1} I\{X_m = i\}] \\
&= \delta_{ij} + \mathbb{E}_j[\sum_{m=1}^{\tau_1} p_{X_{m-1},i}] \\
&= \delta_{ij} + \mathbb{E}_j[\sum_{m=0}^{\tau_1-1} p_{X_m,i}] \\
&= \delta_{ij} + \mathbb{E}_j[\sum_{m=0}^{\tau_1-1} \sum_{k\neq 1} p_{ki} I\{X_m = k\}]
\end{aligned}$$

since $X_m \neq 1$ for $m < \tau_1$,

$$\begin{aligned}
&= \delta_{ij} + \mathbb{E}_j[\sum_{k\neq 1} \sum_{m=0}^{\tau_1-1} p_{ki} I\{X_m = k\}] \\
&= \delta_{ij} + \sum_{k\neq 1} \mathbb{E}_j[\sum_{m=0}^{\tau_1} I\{X_m = k\}] p_{ki},
\end{aligned}$$

since $X_{\tau_1} = 1 \neq k$. Thus,

$$\nu_\lambda^j(i) = \delta_{ij} + \sum_{k\neq 1} \nu_\lambda^j(k) p_{ki}. \tag{3.32}$$

Since $\nu_\lambda^1(i) = 0$ for $i \neq 1$, we also have

$$\nu_\lambda^1(i) = \delta_{i1} + \sum_{k\neq 1} \nu_\lambda^1(k) p_{ki} = 0.$$

Multiplying both sides of (3.32) by $1/N$ and summing over j, we obtain

$$\sum_{j=1}^{N} 1/N \nu_\lambda^j(i) = 1/N + 1/N \sum_j \sum_{k\neq 1} \nu_\lambda^j(k) p_{ki}.$$

This proves the claim. □

Let $J(\lambda)$ denote our objective as a function of λ, that is, w_1^ε, evaluated along the line segment $\{\mathbf{P}_\lambda : 0 \leq \lambda \leq 1\}$. From (3.26), we have

$$J(\lambda) = 1/N^2 \sum_{i=1}^{N} \kappa_\lambda(i). \tag{3.33}$$

Differentiating with respect to λ on both sides yields

$$J'(\lambda) = 1/N^2 \sum_{i=1}^{N} \kappa'_\lambda(i) = 1/N^2 \sum_{i=2}^{N} \kappa'_\lambda(i), \tag{3.34}$$

because $\kappa_\lambda(1) = \mathbb{E}_1[\tau_1] = N$ for all $\lambda \in [0,1]$ and, consequently, $\kappa'_\lambda(1) = 0$ for all $\lambda \in [0,1]$. From (3.28) and the definition of $\tilde{\mathbf{P}}_\lambda$, we have

$$\tilde{\boldsymbol{\kappa}}'_\lambda = (\tilde{\mathbf{P}}_1 - \tilde{\mathbf{P}}_0)\tilde{\boldsymbol{\kappa}}_\lambda + \tilde{\mathbf{P}}_\lambda \tilde{\boldsymbol{\kappa}}'_\lambda.$$

Therefore,

$$\tilde{\boldsymbol{\kappa}}'_\lambda = (\mathbf{I} - \tilde{\mathbf{P}}_\lambda)^{-1}(\tilde{\mathbf{P}}_1 - \tilde{\mathbf{P}}_0)\tilde{\boldsymbol{\kappa}}_\lambda,$$

and, together with (3.29) and (3.30), this leads to

$$\begin{aligned} J'(\lambda) &= \frac{1}{N^2}\tilde{\mathbf{e}}^{\mathrm{T}}(\mathbf{I} - \tilde{\mathbf{P}}_\lambda)^{-1}(\tilde{\mathbf{P}}_1 - \tilde{\mathbf{P}}_0)\tilde{\boldsymbol{\kappa}}_\lambda \\ &= \frac{1}{N}\tilde{\boldsymbol{\nu}}_\lambda^{\mathrm{T}}(\tilde{\mathbf{P}}_1 - \tilde{\mathbf{P}}_0)\tilde{\boldsymbol{\kappa}}_\lambda \\ &= \frac{1}{N^2}\tilde{\mathbf{e}}^{\mathrm{T}}(\mathbf{I} - \tilde{\mathbf{P}}_\lambda)^{-1}(\tilde{\mathbf{P}}_1 - \tilde{\mathbf{P}}_0)(\mathbf{I} - \tilde{\mathbf{P}}_\lambda)^{-1}\tilde{\mathbf{e}}. \end{aligned} \tag{3.35}$$

Though the following, purely technical, lemma is a straightforward application of the Cauchy-Schwartz inequality, we include its proof for the sake of completeness.

Lemma 3.4. *If $x_m = m$ for $1 \le m \le N$, and $\{y_k\}$ is a permutation of $\{x_j\}$, then $\sum_i x_i y_i$ is maximised when $y_i = x_i$ for all i, and minimised when $y_i = N + 1 - x_i$ for all i.*

Proof. The maximisation claim is immediate from the Cauchy-Schwartz inequality. For the permutation $\{z_i = N + 1 - y_i\}$ of $\{x_j\}$, we have

$$\sum_i x_i z_i \le \sum_i x_i^2,$$

with equality if and only if $x_i = z_i$ for all i. Hence,

$$\sum_i x_i y_i = \sum_i x_i(N + 1 - z_i) \ge \sum_i x_i(N + 1 - x_i),$$

with equality if and only if $y_i = N + 1 - x_i$ for all i. Thus, the minimisation claim also follows. □

We now consider $J'(0)$ in the situations where the doubly stochastic matrix $\mathbf{P}_0 \in \mathcal{D}_{\text{H}}$ is induced by a deterministic transition matrix tracing out a Hamiltonian cycle. We first show that $J'(0) > 0$ on a straight line path from $\mathbf{P}_0$ towards any doubly stochastic $\mathbf{P}_1$ induced by the graph. This shows that deterministic transition matrices inducing Hamiltonian cycles correspond to local minima. Suppose then that $\mathbf{P}_0$ corresponds to a Hamiltonian cycle H_0. Without loss of generality, we assume H_0 is the Hamiltonian cycle $1 \to 2 \to \cdots \to N \to 1$. To start with, we consider $\mathbf{P}_1 \in \mathcal{D}_{\text{H}} \cup \mathcal{D}_s$, other than $\mathbf{P}_0$. That is, $\mathbf{P}_1$ is induced by any deterministic matrix that traces out in the graph either a union of disjoint cycles or a Hamiltonian cycle other than H_0. For each $i \in \mathcal{S}$ and $i \neq 1$, we denote by $m(i)$ the number of steps required to reach vertex 1 from i on H_0 if $\varepsilon = 0$. Then,

$$\begin{aligned} \kappa_0(i) &= \mathbb{E}_i[\tau_1] = m(i) + O(\varepsilon) \\ &= (N - i + 1) + O(\varepsilon). \end{aligned} \tag{3.36}$$

To verify that the suppressed $\varepsilon-$dependence above is, indeed, only of order $O(\varepsilon)$, note that by (3.13) we can write $\tilde{\mathbf{P}}_0 = \tilde{\mathbf{P}}_0(\varepsilon) = \tilde{\mathbf{P}}_0(0) - \varepsilon\tilde{\mathbf{K}}_0$, for some fixed matrix $\tilde{\mathbf{K}}_0$. Now, it follows from (3.29) that (3.36) is simply the ith equation in the system

$$\tilde{\boldsymbol{\kappa}}_0 = (\mathbf{I} - \tilde{\mathbf{P}}_0(\varepsilon))^{-1}\tilde{\mathbf{1}} = (\mathbf{I} - \tilde{\mathbf{P}}_0(0) + \varepsilon\tilde{\mathbf{K}}_0)^{-1}\tilde{\mathbf{1}}. \tag{3.37}$$

It is easy to check that $\mathbf{I} - \tilde{\mathbf{P}}_0(0)$ is invertible and hence that the above equation corresponds to a regularly (rather than a singularly) perturbed system. By well-known results from perturbation theory of linear operators (see Langenhop [73] and Avrachenkov *et al.* [6]), it follows that

$$\tilde{\boldsymbol{\kappa}}_0 = (\mathbf{I} - \tilde{\mathbf{P}}_0(0))^{-1}\tilde{\mathbf{1}} + O(\varepsilon).$$

Also,

$$\mathbb{E}_j[\sum_{\ell=0}^{\tau_1} I\{X_\ell = i\}] = \begin{cases} 1 + O(\varepsilon) & \text{for } j = 1, 2, \ldots, i-1, \\ O(\varepsilon) & \text{for } j = i, \ldots, N. \end{cases} \tag{3.38}$$

Equation (3.38) can be proved by arguments analogous to those used for proving (3.36). Hence, for $i = 2, \ldots, N$,

$$\begin{aligned} \nu_0(i) &= 1/N \sum_{1 \le j < i} \mathbb{E}_j[\sum_{\ell=1}^{\tau_1} I\{X_\ell = i\}] + O(\varepsilon) \\ &= (i-1)/N + O(\varepsilon). \end{aligned}$$

Thus, by (3.28),

$$
\begin{aligned}
\tilde{\boldsymbol{\nu}}_0^{\mathrm{T}}\tilde{\mathbf{P}}_0\tilde{\boldsymbol{\kappa}}_0 &= \sum_{i=2}^{N-1} \nu_0(i)\kappa_0(i+1) \\
&= 1/N\{1(N-2)+2(N-3)+\cdots(N-2)1\}+O(\varepsilon) \\
&= 1/N\sum_{r=1}^{N-1} r\{(N-1)-r\}+O(\varepsilon) \\
&= (N-1)^2/2-1/N\sum_{r=1}^{N-1} r^2+O(\varepsilon).
\end{aligned}
$$

Now, suppose that $\mathbf{P}_1$ is induced by either a Hamiltonian cycle distinct from H_0 or a deterministic matrix that traces out a union of disjoint cycles in the graph. Hence, for every ith row there is a unique j_ith column such that $[\mathbf{P}_1]_{i,j_i} = 1$ and $j_i \neq j_k$, for $i \neq k$. Thus,

$$
\begin{aligned}
\tilde{\boldsymbol{\nu}}_0^{\mathrm{T}}\tilde{\mathbf{P}}_1\tilde{\boldsymbol{\kappa}}_0 &= \sum_{i=2}^{N} \nu_0(i)\kappa_0(j_i) \\
&= 1/N\sum_{i=2}^{N}(i-1)(N-j_i+1)+O(\varepsilon) \\
&= 1/N\sum_{r=1}^{N-1} r\{(N-1)-(j_{r+1}-2)\}+O(\varepsilon) \\
&= (N-1)/N\sum_{r=1}^{N-1} r-1/N\sum_{r=1}^{N-1} ry_r+O(\varepsilon) \\
&= (N-1)^2/2-1/N\sum_{r=1}^{N-1} ry_r+O(\varepsilon),
\end{aligned}
$$

where $r = i-1$, $y_r = (j_{r+1}-2)$ and $y_r \in \{0,1,2,\ldots,(N-2)\}$ with $y_r \neq y_k$ whenever $r \neq k$. If y_r were allowed to take values only in the set $\{1,2,\ldots,N-1\}$, then by Lemma 3.4 we would have that

$$
\sum_{r=1}^{N-1} r^2 > \sum_{r=1}^{N-1} ry_r, \tag{3.39}
$$

whenever $(y_1,\ldots,y_{N-1}) \neq (1,\ldots,N-1)$. However, the inclusion of zero as one of the possible values for y_r can only lower the right-hand side of (3.39). Hence, we have proved that, whenever $\tilde{\mathbf{P}}_1 \in \mathcal{D}_{\mathrm{H}} \cup \mathcal{D}_s$ and $\tilde{\mathbf{P}}_1 \neq \tilde{\mathbf{P}}_0$,

$$
\tilde{\boldsymbol{\nu}}_0^{\mathrm{T}}\tilde{\mathbf{P}}_1\tilde{\boldsymbol{\kappa}}_0 - \tilde{\boldsymbol{\nu}}_0^{\mathrm{T}}\tilde{\mathbf{P}}_0\tilde{\boldsymbol{\kappa}}_0 > 0.
$$

Now consider an arbitrary doubly stochastic $\mathbf{P}_1$ other than $\mathbf{P}_0$. By Birkhoff-von Neumann theorem (Bapat and Raghavan [9]),

$$\mathbf{P}_1 = \sum_{i=1}^{M} \gamma_i \mathbf{P}_i^{\dagger}, \tag{3.40}$$

where $\gamma_i \geq 0$ for all $i, \sum_i \gamma_i = 1$, $\mathbf{P}_i^{\dagger} \in \mathcal{D}_{\text{H}} \cup \mathcal{D}_s$ correspond to permutation matrices and $M \geq 1$ is the number of permutation matrices induced by the graph. For at least one value of i in the summation (3.40), $\mathbf{P}_i^{\dagger} \neq \mathbf{P}_0$ and $\gamma_i > 0$. Then, by the preceding strict inequalities and the second equality of (3.35) we have that

$$\begin{aligned} J'(0) &= 1/N(\tilde{\boldsymbol{\nu}}_0^{\mathrm{T}} \tilde{\mathbf{P}}_1 \tilde{\boldsymbol{\kappa}}_0 - \tilde{\boldsymbol{\nu}}_0^{\mathrm{T}} \tilde{\mathbf{P}}_0 \tilde{\boldsymbol{\kappa}}_0) \\ &= 1/N \sum_i \gamma_i (\tilde{\boldsymbol{\nu}}_0^{\mathrm{T}} \tilde{\mathbf{P}}_i^{\dagger} \tilde{\boldsymbol{\kappa}}_0 - \tilde{\boldsymbol{\nu}}_0^{\mathrm{T}} \tilde{\mathbf{P}}_0 \tilde{\boldsymbol{\kappa}}_0) > 0. \end{aligned}$$

The following main result now follows rather easily.

Theorem 3.5. *If* $\mathbf{P}_0$ *is induced by a Hamiltonian cycle, then,*

(i) $\mathbf{P}_0$ *is a strict local minimum for the cost functional* w_1^{ε}*, and*

(ii) $\mathbf{P}_0$ *is also a global minimum for the cost functional* w_1^{ε}*.*

Proof. Part (i) was proved above for $\mathbf{P}_0$ corresponding to the Hamiltonian cycle H_0: it is sufficient to observe that for a strict *local* minimum, the quantity

$$\boldsymbol{\nu}_0 \tilde{\mathbf{P}}_1 \boldsymbol{\kappa}_0 - \boldsymbol{\nu}_0 \tilde{\mathbf{P}}_0 \boldsymbol{\kappa}_0$$

remains strictly bounded away from zero as ε approaches 0 for all extremal $\mathbf{P}_1 \neq \mathbf{P}_0$. The effect of considering another Hamiltonian cycle would be only to permute the order of the terms in various summations, without changing the conclusions.

To obtain Part (ii), first note that the above allows us to choose an $\eta_0 > 0$ such that $\mathbf{P}_0$ is the strict local minimum of w_1^{ε} in the η_0-neighborhood of $\mathbf{P}_0$. As in the proof of Theorem 3.4, choose $\varepsilon > 0$ small enough so that the global minimum of w_1^{ε} is attained on the η_0-neighborhood of $\mathbf{P}_0$. *Small enough* here is quantified by an upper bound that depends only on N and η_0 (see Borkar *et al.* [17]). The claim follows. □

Recall from Theorem 3.3 that for a doubly stochastic matrix $\mathbf{P} \in \mathcal{D}$, the functional consisting of the top-left element of the fundamental matrix induced by $\mathbf{P}$ is given by

$$w_1^{\varepsilon} = \frac{N+1}{2N} + \frac{1}{2N^2} \mathbb{E}_1[(\tau_1 - N)^2]. \tag{3.41}$$

We suppressed the dependence on $\mathbf{P}$ on the right-hand side of (3.41), but the expectation term is a function of $\mathbf{P}$, since $\mathbf{P}$ determines the distribution of τ_1. It should now be clear that a consequence of Theorems 3.3 and 3.5 is that whenever the underlying graph G is Hamiltonian, the minimum of the above functional over $\mathbf{P} \in \mathcal{D}$ is given by

$$w_1^\varepsilon(\mathbf{P}_{\mathrm{H}}) = \min_{\mathbf{P} \in \mathcal{D}} w_1^\varepsilon(\mathbf{P}) = (N+1)/(2N) + O(\varepsilon), \tag{3.42}$$

where $\mathbf{P}_{\mathrm{H}} \in \mathcal{D}_{\mathrm{H}}$ is a probability transition matrix defining any Hamiltonian cycle in the graph.

Hamiltonicity Gap: A lower bound for the non-Hamiltonian case

In this section, we prove that for positive and sufficiently small ε, there exists $\Delta(N) > 0$ such that whenever the graph G is non-Hamiltonian

$$\{\min_{\mathbf{P} \in \mathcal{D}^\varepsilon} w_1^\varepsilon(\mathbf{P})\} - w_1^\varepsilon(\mathbf{P}_{\mathrm{H}}) \geq \Delta(N) - O(\varepsilon).$$

We name the quantity $\Delta(N)$ the *Hamiltonicity gap of order* N because it distinguishes all non-Hamiltonian graphs with N vertices from all Hamiltonian graphs with the same number of vertices.

Before presenting the proof, we note that such a result is reasonable when we consider the possible variability of τ_1—as captured by its variance $\mathbb{E}_1[(\tau_1 - N)^2]$—for both Hamiltonian and non-Hamiltonian graphs. In the former case, it is clear that this variance can be made nearly zero by following a Hamiltonian cycle because the latter would yield a variance actually equal to zero were it not for the (small) perturbation ε. However, if the graph is non-Hamiltonian, perhaps we cannot avail ourselves of such a variance annihilating transition matrix. This intuitive reasoning is made rigorous in the remainder of this section.

The key step in what follows is the derivation of an upper bound on $P\{\tau_1 = N|\mathbf{P}\}$, the probability that the system returns to vertex 1 in N steps, under an arbitrary doubly stochastic matrix $\mathbf{P}$ in a non-Hamiltonian graph.

Lemma 3.5. *Suppose that G is a non-Hamiltonian graph, and let $\mathbf{P}$ be an arbitrary doubly stochastic transition matrix feasible on G.*

(i) If $\varepsilon = 0$, then $P\{\tau_1 = N|\mathbf{P}\} \leq 1/4$.

(ii) If $\varepsilon > 0$ and small, then $P\{\tau_1 = N|\mathbf{P}\} \leq 1/4 + O(\varepsilon)$.

Proof. First, consider the case $\varepsilon = 0$. Let $\mathbf{P}$ be an arbitrary doubly stochastic matrix and let $\{X_t\}_0^\infty$ be the Markov chain induced by $\mathbf{P}$ and the starting state 1. Let $\gamma_1 = (X_0, X_1, \ldots, X_N)$ be a path of N steps through the graph and let $\chi_1 = \{\gamma_1 | X_0 = X_N = 1, X_k \neq 1, k = 1, \ldots, N-1\}$. That is, the event that the first return to 1 occurs after N steps is $\{\tau_1 = N\}$, which is simply the event that γ_1 traces a path within χ_1 and hence

$$P\{\tau_1 = N|\mathbf{P}\} = \sum_{\gamma_1 \in \chi_1} p_{\gamma_1},$$

where p_{γ_1} denotes the probability (under $\mathbf{P}$) of observing the path γ_1. However, because the graph is assumed to be non-Hamiltonian, all the paths in χ_1 that receive a positive probability have the structure

$$\gamma_1 = \gamma_1' \cup \bar{\gamma}_1,$$

where γ_1' consists of a non-self-intersecting "reduced path" from 1 to itself of length $m \leq N-2$ adjoined at some vertex (or vertices) other than 1 by one or more loops of total length $N-m$, that together constitute $\bar{\gamma}_1$. One can think of $\gamma_1' \cup \bar{\gamma}_1$ as the first and second parts of a figure comprising of a basic loop with one or more side-lobes attached to it, each of which is either a loop or a connected union of loops. The simplest instance of this is a figure of eight, with two loops of length m and $N-m$ respectively, attached at a vertex other than 1.

Let p_{γ_1} denote the probability of the original path and p'_{γ_1} that of the reduced path. Let $q = p_{\gamma_1}/p'_{\gamma_1} \leq 1$, which is the contribution to p coming from the loops comprising $\bar{\gamma}_1$. More generally, define $\gamma_0 = \gamma_1', \gamma_1 = \gamma_1' \cup \bar{\gamma}_1, \gamma_2 = \gamma_1' \cup \bar{\gamma}_1 \cup \bar{\gamma}_1, \gamma_3 = \gamma_1' \cup \bar{\gamma}_1 \cup \bar{\gamma}_1 \cup \bar{\gamma}_1, \ldots$. For $n \geq 2$, the paths γ_n from 1 to itself that begin with the same reduced path γ_1' but may repeat exactly the path $\bar{\gamma}_1$ for $n \geq 2$ times, all contribute to the event $\{\tau_1 \neq N\}$, as does $\gamma_0 = \gamma_1'$.

The paths γ_n, for $n \geq 2$, have probabilities $p_{\gamma_1} q^{n-1}$. The total probability that these paths and $\gamma_0 = \gamma_1'$ (but excluding the original γ_1) contribute to $\{\tau_1 \neq N\}$ is

$$\begin{aligned} p_{\gamma_1}/q + \sum_{n \geq 2} p_{\gamma_1} q^{n-1} &= p_{\gamma_1}\{1/q + q/(1-q)\} \\ &= p_{\gamma_1}\{-1 + 1/(q(1-q))\} \\ &\geq 3p_{\gamma_1}. \end{aligned}$$

It follows that

$$P\{\tau_1 \neq N|\mathbf{P}\} \geq \sum_{\gamma_1 \in \chi_1} 3p_{\gamma_1} = 3P\{\tau_1 = N|\mathbf{P}\}.$$

Hence,

$$\begin{aligned} 1 &= P\{\tau_1 < \infty|\mathbf{P}\} \\ &= P\{\tau_1 = N|\mathbf{P}\} + P\{\tau_1 \neq N|\mathbf{P}\} \\ &\geq 4P\{\tau_1 = N|\mathbf{P}\}, \end{aligned}$$

implying $P\{\tau_1 = N|\mathbf{P}\} \leq 1/4$, or, $P\{\tau_1 \neq N|\mathbf{P}\} \geq 3/4$.

Returning to the case when $\varepsilon > 0$ and sufficiently small, we note that in the Markov chain induced by $\mathbf{P}$ there are now two types of transitions: *strong* transitions that correspond to $\mathbf{P}$ assigning a positive probability to edges that are actually in the graph and *weak* transitions that are strictly the result of our perturbation. The latter are of order ε. Thus, the only impact that the perturbation makes on the argument presented above is to introduce an adjustment of order ε. This completes the proof. □

Theorem 3.6. *Consider a non-Hamiltonian graph G of order N, and define $\Delta(N) = 3/(8N^2)$.*

(i) For any $\mathbf{P}$, $\mathbb{E}_1[(\tau_1 - N)^2] \geq 3/4 - O(\varepsilon)$.

(ii) The following lower bound holds:

$$\{\min_{\mathbf{P}\in\mathcal{D}^\varepsilon} w_1^\varepsilon(\mathbf{P})\} - w_1^\varepsilon(\mathbf{P}_{\text{H}}) \geq \Delta(N) - O(\varepsilon).$$

Proof. Let $\mathbf{P}$ be an arbitrary doubly stochastic matrix and $\mathbb{E}_1[(\tau_1 - N)^2]$ be the corresponding variance of the first return time to vertex 1, starting from 1. Clearly,

$$\begin{aligned} \mathbb{E}_1[(\tau_1 - N)^2] &= \sum_{k\geq 1}(k-N)^2 P\{\tau_1 = k|\mathbf{P}\} \\ &\geq \sum_{k\geq 1,\ k\neq N} P\{\tau_1 = k|\mathbf{P}\} \\ &= P\{\tau_1 \neq N|\mathbf{P}\}. \end{aligned}$$

Hence by Part (ii) of Lemma 3.5, we have obtained Part (i), namely

$$\mathbb{E}_1[(\tau_1 - N)^2] \geq 3/4 - O(\varepsilon). \tag{3.43}$$

It now follows from (3.41) that

$$u_1^\varepsilon \geq \frac{N+1}{2N} + \frac{1}{2N^2}(3/4 - O(\varepsilon)) = \frac{N+1}{2N} + \Delta(N) - O(\varepsilon). \tag{3.44}$$

Part (ii) now follows immediately from (3.44) and (3.42). □

In summary, in this chapter, we considered spaces of probability transition matrices of Markov chains induced by a given graph G and certain associated random variables such as first return times to given vertices. We showed that, in a prescribed sense, the variances of the latter are minimised precisely at those Markov chains that are induced by Hamiltonian cycles of the graph whenever it possesses such cycles. Furthermore, we showed that the perturbed variance functional differentiates between all Hamiltonian and non-Hamiltonian graphs of a given order, by means of the size of the gap between minimal values of that functional over the space of doubly stochastic probability transition matrices induced by G. This suggests that stochastic—perhaps even statistical—methods could be brought to bear on this essentially deterministic, combinatorial, problem.

Chapter 4
Markov Decision Processes

4.1 Introduction

Markov chains are useful in describing many discrete event stochastic processes; however, they are not flexible enough to model situations where we have to make decisions to control the future trajectories of the system. For this reason, the theory of *Markov decision processes* (MDPs), also known as *controlled Markov chains*, has been developed. In particular, in the context of this book, we observe that in any given graph Hamiltonian cycles (if any) correspond to a family of spanning subgraphs inducing special Markov chains whose probability transition matrices are a subset of permutation matrices possessing only a single ergodic class. However, the family of all probability transition matrices naturally induced by the same graph constitutes a much richer, convex, domain. Markov decision processes provide a well developed body of techniques for searching both this domain and associated, also convex, domains of occupational measures induced by stationary policies of MDPs. In this chapter we explore the structure of one of these domains in more detail and identify characteristics of its extreme points that lend themselves to algorithmic developments described in subsequent chapters.

4.2 Markov Decision Processes

A Markov decision process is similar to a Markov chain, except that the transition probabilities—the probabilities of the process going from one state to another within the state space—depend on the *actions* taken by a *controller*. At each step of transition, we receive a certain value or *reward*. The objective then is to find an optimal policy that specifies a sequence of deterministic or randomised actions in order to maximise the reward obtained. The theory of MDPs is useful in modelling certain real-life tasks involving planning under uncertainty, such as scheduling, robotics and financial decision making (see Puterman [87] for an excellent introduction to MDPs).

Action Spaces In an MDP, in addition to the state space $\mathcal{S}$, we have a set of *action spaces*. For each $i \in \mathcal{S}$, an *action space* $\mathcal{A}(i)$ is a finite set of actions that the controller may choose in state i. We assume that $\mathcal{A}(i) \neq \emptyset$ for all $i \in \mathcal{S}$. The *total action space* $\mathcal{A}$ then is defined as the union of all $\mathcal{A}(i)$, that is,

$$\mathcal{A} = \bigcup_{i=1}^{N} \mathcal{A}(i). \tag{4.1}$$

Probability Transition Law It is assumed that the stationary transition probabilities $p(j|i,a)$ are known and satisfy

$$\sum_{j \in \mathcal{S}} p(j|i,a) = 1, \quad a \in \mathcal{A}(i), i \in \mathcal{S}. \tag{4.2}$$

The $p(j|i,a)$'s denote the probabilities of the system moving from state i to state j, given that action $a \in \mathcal{A}(i)$ is selected in state i, independently of when the latter state is visited. These probabilities form an essential part of the data defining a Markov decision process.

Policies and Reward Functions As from each $i \in \mathcal{S}$ there is one or more states to which the system can move, a selection of a *policy* $\boldsymbol{f}$ determines which state(s) it will visit and with what probability.

If the probability of a policy choosing a particular action a does not depend on time t or any previous state or action, other than through the current state, then we have stationary transition probability. Let A_t be the action chosen at time t. A policy is *stationary* if all decisions are only dependent on the current state i, and not the time t[1], that is, for $a \in \mathcal{A}(i)$, $i \in \mathcal{S}$ and $t = 0, 1, 2, \ldots,$

$$P\{A_t = a | X_t = i\} = f(i,a) \text{ for some } \boldsymbol{f} : \bigcup_{i \in \mathcal{S}} \{i\} \times \mathcal{A}(i) \to [0,1].$$

It is important to note that each stationary policy $\boldsymbol{f}$ uniquely defines a Markov chain. Let $\mathcal{F}_\mathcal{S}$ be the set of all stationary policies in an MDP. A policy $\boldsymbol{f} \in \mathcal{F}_\mathcal{S}$ is *deterministic* if at each state $i \in \mathcal{S}$, it chooses exactly one $a \in \mathcal{A}(i)$. In this case, $f(i,a) \in \{0,1\}$ for all $i \in \mathcal{S}, a \in \mathcal{A}(i)$. We denote by $\mathcal{F}_\mathcal{D}$ the set of all deterministic policies in an MDP, $\mathcal{F}_\mathcal{D} \subset \mathcal{F}_\mathcal{S}$. On the other

---

[1] In Markov decision processes, it is customary to also consider more general policies that may depend on histories of the process. However, in our application of MDPs to the HCP, only stationary policies will be needed.

hand, a policy is *randomised* if for at least one state $i \in \mathcal{S}$, it chooses two or more $a \in \mathcal{A}(i)$. In both cases, $f(i,a) \in [0,1]$ and for every $i \in \mathcal{S}$,

$$\sum_{a \in \mathcal{A}(i)} f(i,a) = 1. \tag{4.3}$$

Every MDP also has a *reward function* $r(i,a)$, which determines the reward given to the policy for choosing $a \in \mathcal{A}(i)$ at state i. We define $\boldsymbol{r}(\boldsymbol{f}) = (r_1(\boldsymbol{f}), \ldots, r_N(\boldsymbol{f}))^{\mathrm{T}}$, where $r_i(\boldsymbol{f}) = \sum_{a \in \mathcal{A}(i)} r(i,a) f(i,a)$.

Probability Transition Matrix Associated with every $\boldsymbol{f} \in \mathcal{F}_{\mathcal{S}}$, we define a matrix $\mathbf{P}(\boldsymbol{f}) = [p_{ij}]$ where, for $i, j = 1, \ldots, N$,

$$p_{ij} = P(j|i, \boldsymbol{f}) = \sum_{a \in \mathcal{A}(i)} p(j|i,a) f(i,a). \tag{4.4}$$

The matrix $\mathbf{P}(\boldsymbol{f})$ is the probability transition matrix of the Markov chain induced by $\boldsymbol{f}$. Similarly to (3.5), define the stationary distribution matrix $\mathbf{P}^*(\boldsymbol{f})$ as follows

$$\mathbf{P}^*(\boldsymbol{f}) = \lim_{T \to \infty} \frac{1}{T+1} \sum_{t=0}^{T} \mathbf{P}^t(\boldsymbol{f}). \tag{4.5}$$

Fundamental Matrix For every Markov chain, the matrix $\mathbf{I} - \mathbf{P}(\boldsymbol{f}) + \mathbf{P}^*(\boldsymbol{f})$ is always invertible, and its inverse is called the *fundamental matrix*. Let $\mathbf{G}(\boldsymbol{f})$ be the fundamental matrix of a controlled Markov chain induced by $\boldsymbol{f}$, namely

$$\mathbf{G}(\boldsymbol{f}) = \{\mathbf{I} - \mathbf{P}(\boldsymbol{f}) + \mathbf{P}^*(\boldsymbol{f})\}^{-1}. \tag{4.6}$$

Typically, we are interested in finding one or more policies $\boldsymbol{f}$ that is optimal according to a suitable criterion, over a specified time horizon.

Evaluation of Rewards Let R_t be the reward for the period $[t, t+1)$. For randomised policies, we have an *expected reward*

$$\begin{aligned} \mathbb{E}_{i\boldsymbol{f}}[R_t] &= \mathbb{E}_{\boldsymbol{f}}[R_t | S_0 = i] \\ &= [\mathbf{P}^t(\boldsymbol{f}) \boldsymbol{r}(\boldsymbol{f})]_i, \qquad \text{for } t = 0, 1, 2, \ldots \end{aligned} \tag{4.7}$$

which is dependent on the selected policy $\boldsymbol{f}$ and the initial state S_0. There are three standard ways of evaluating this sequence of expected rewards:

1. We consider finite horizon situations, where there are a finite number of steps, and we are able to evaluate the total expected reward.

2. We calculate the average of the accumulated rewards, over a finite time horizon, and let the length of that horizon tend to infinity. This procedure

defines the *limiting average reward*.

3. We consider a *discounted model*, where each successive reward is further discounted by a multiplier lying between 0 and 1, so that earlier rewards carry more weight than the later ones. This multiplier is called a *discount factor*.

We briefly describe the last two options.

Limiting Average Process For a stationary policy $\boldsymbol{f}$, with the initial distribution vector $\boldsymbol{\nu} = \boldsymbol{e}_i^{\mathrm{T}}$, where $\boldsymbol{e}_i^{\mathrm{T}} = (0, \ldots, 1, \ldots, 0)$ with 1 being at the ith position, the *limiting average value* is defined as

$$v(\boldsymbol{e}_i^{\mathrm{T}}, \boldsymbol{f}) = \lim_{T \to \infty} \frac{1}{T+1} \sum_{t=0}^{T} \mathbb{E}_{i\boldsymbol{f}}[R_t]. \tag{4.8}$$

Consequently, we have the *limiting average value vector* of $\boldsymbol{f}$

$$\boldsymbol{v}(\boldsymbol{f}) = (v(\boldsymbol{e}_1^{\mathrm{T}}, \boldsymbol{f}), v(\boldsymbol{e}_2^{\mathrm{T}}, \boldsymbol{f}), \ldots, v(\boldsymbol{e}_N^{\mathrm{T}}, \boldsymbol{f}))^{\mathrm{T}}.$$

We say that $\boldsymbol{v}(\boldsymbol{f}_1) \geq \boldsymbol{v}(\boldsymbol{f}_2)$ if $v(\boldsymbol{e}_i^{\mathrm{T}}, \boldsymbol{f}_1) \geq v(\boldsymbol{e}_i^{\mathrm{T}}, \boldsymbol{f}_2)$ for every $i \in \mathcal{S}$. The objective of our optimisation problem is to find an optimal control $\boldsymbol{f}^0$ such that $\boldsymbol{v}(\boldsymbol{f}^0)$ is maximum, that is,

$$\boldsymbol{v}(\boldsymbol{f}^0) = \max_{\boldsymbol{f}} \boldsymbol{v}(\boldsymbol{f}).$$

From (4.7) and (4.8), it now follows that

$$\begin{aligned} \boldsymbol{v}(\boldsymbol{f}) &= \lim_{T \to \infty} \frac{1}{T+1} \sum_{t=0}^{T} \mathbf{P}^t(\boldsymbol{f}) \boldsymbol{r}(\boldsymbol{f}) \\ &= \{\lim_{t \to \infty} \mathbf{P}^t(\boldsymbol{f})\} \boldsymbol{r}(\boldsymbol{f}) \\ &= \mathbf{P}^*(\boldsymbol{f}) \boldsymbol{r}(\boldsymbol{f}). \end{aligned} \tag{4.9}$$

A *limiting average Markov decision process* is a model that employs the above limiting average value vector as a criterion for evaluating performance of different policies.

Discounted Model For a stationary policy $\boldsymbol{f}$ with the initial distribution vector $\boldsymbol{v} = \boldsymbol{e}_i^{\mathrm{T}}$, the *discounted value* is defined as

$$v_\beta(\boldsymbol{e}_i^{\mathrm{T}}, \boldsymbol{f}) = \sum_{t=0}^{\infty} \beta^t \mathbb{E}_{i\boldsymbol{f}}[R_t],$$

where $\beta \in (0,1)$ is the *discount factor*. Consequently, from (4.7), we have the *discounted value vector* of $\boldsymbol{f}$

$$\begin{aligned}\boldsymbol{v}_\beta(\boldsymbol{f}) &= (v_\beta(\boldsymbol{e}_1^{\mathrm{T}},\boldsymbol{f}), v_\beta(\boldsymbol{e}_2^{\mathrm{T}},\boldsymbol{f}),\ldots,v_\beta(\boldsymbol{e}_N^{\mathrm{T}},\boldsymbol{f}))^{\mathrm{T}}\\ &= \sum_{t=0}^{\infty}\beta^t\mathbf{P}^t(\boldsymbol{f})\boldsymbol{r}(\boldsymbol{f})\\ &= [\mathbf{I}-\beta\mathbf{P}(\boldsymbol{f})]^{-1}\boldsymbol{r}(\boldsymbol{f}), \end{aligned} \tag{4.10}$$

where $\mathbf{P}^0(\boldsymbol{f})$ is the $N \times N$ identity matrix, and the last equality employs a well-known result that the matrix $\mathbf{I}-\beta\mathbf{P}(\boldsymbol{f})$ is invertible for every stationary policy $\boldsymbol{f}$. Furthermore, with the initial distribution vector $\boldsymbol{\nu}^{(0)} = \boldsymbol{e}_i^{\mathrm{T}}$, the discounted value is

$$v_\beta(\boldsymbol{e}_i^{\mathrm{T}},\boldsymbol{f}) = \{[\mathbf{I}-\beta\mathbf{P}(\boldsymbol{f})]^{-1}\boldsymbol{r}(\boldsymbol{f})\}_i. \tag{4.11}$$

We say $\boldsymbol{v}_\beta(\boldsymbol{f}_1) \geq \boldsymbol{v}_\beta(\boldsymbol{f}_2)$ if $v_\beta(\boldsymbol{e}_i^{\mathrm{T}},\boldsymbol{f}_1) \geq v_\beta(\boldsymbol{e}_i^{\mathrm{T}},\boldsymbol{f}_2)$ for every $i \in \mathcal{S}$. The objective of our optimisation problem, similar to that of the limiting average process, is to find an optimal control $\boldsymbol{f}^0$ such that $\boldsymbol{v}_\beta(\boldsymbol{f}^0)$ is maximum, that is,

$$\boldsymbol{v}_\beta(\boldsymbol{f}^0) = \max_{\boldsymbol{f}} \boldsymbol{v}_\beta(\boldsymbol{f}).$$

For both of these models, there exists a deterministic policy that gives the optimal value for the associated objective function [12].

Embedding of a graph in a Markov decision process Begining with a given graph $G = (V(G), E(G))$, we embed this graph in a Markov decision process by setting the state space $\mathcal{S} = V(G)$, the set of vertices of G. The action set $\mathcal{A}(i) = \{j \in \mathcal{S} | (i,j) \in E(G)\}$ is the set of vertices that can be reached from $i \in \mathcal{S}$ in one step in the graph G. It will also be useful to introduce $\mathcal{B}(i) = \{j \in \mathcal{S} | (j,i) \in E(G)\}$, the set of vertices from which the system can travel to i in one step in the graph G. The probability transition law of the Markov decision process is defined by

$$p(j|i,a) = \begin{cases} 1 & \text{if } a \in \mathcal{A}(i) \text{ and } j = a,\\ 0 & \text{otherwise,}\end{cases} \tag{4.12}$$

for all $a \in \mathcal{A}(i)$ and $i, j \in \mathcal{S}$. Note that (4.12) is very natural because it merely states that the system will travel from i to j if and only if, at i, the decision maker chooses an edge that has j as the other vertex.

We specify the reward vector $\boldsymbol{r}(\boldsymbol{f})$ formally as an N-dimensional column vector, the ith element of which is the immediate expected reward for being at vertex $i \in \mathcal{S}$. As any Hamiltonian cycle visits every vertex exactly once before returning to the beginning, it is important to identify where the system starts and when it revisits that vertex. While we need to know whether the system originates at a single vertex, the label of the vertex is unimportant because of symmetry. For these reasons, we always start the system at vertex 1, which we call the *home vertex*; hence, the default N-dimensional initial distribution vector is $\boldsymbol{\nu}^{(0)} = \boldsymbol{e}_1^{\mathrm{T}} = (1, 0, \ldots, 0)$. Using $\boldsymbol{\nu}^{(0)}$, we decompose the set of all deterministic policies feasible for a given graph into three exhaustive and mutually exclusive subsets:

- *noose cycles*—non-Hamiltonian deterministic policies for which the home vertex 1 is transient

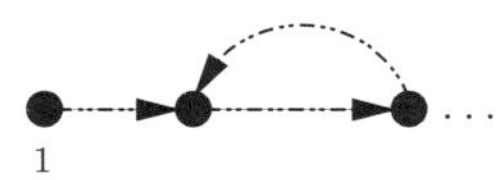

Fig. 4.1: A noose cycle

- *short cycles*—non-Hamiltonian deterministic policies for which the home vertex 1 is recurrent

Fig. 4.2: A short cycle.

- *Hamiltonian cycles.*

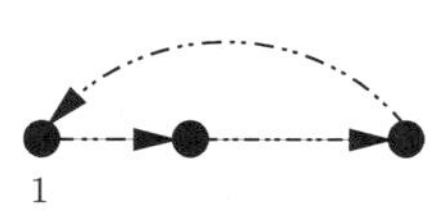

Fig. 4.3: A Hamiltonian cycle

By this classification, in a noose cycle, the home vertex is not a part of any cycle, and in a short cycle, it is part of a cycle that has a length less than N. In other words, after starting at the home vertex, the first vertex a short cycle revisits is the home vertex itself, whereas the first vertex a noose cycle revisits is *not* the home vertex. In fact, a noose cycle can never return to 1 after leaving it.

For every policy $\boldsymbol{f}$, let $\boldsymbol{r}(\boldsymbol{f}) = \boldsymbol{r} = (1, 0, \ldots, 0)^{\mathrm{T}}$ to distinguish the home vertex from any other vertex. Additionally, the discounted model assists us in recognising when the system revisits the home vertex at the time t, as the reward received at the tth step is assigned an unique coefficient β^t, or 0 if the vertex visited at that step is not the home vertex.

Example 4.1 *Consider the 4-vertex complete graph, depicted in Figure 4.4.*

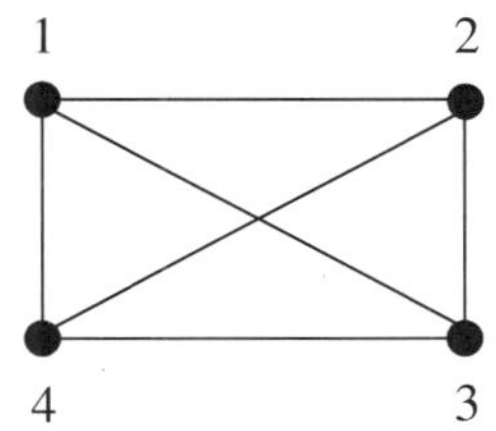

Fig. 4.4: The 4-vertex complete graph

The policy $\boldsymbol{f}_{\mathrm{H}} : \{1,2,3,4\} \to \{2,3,4,1\}$ *is a Hamiltonian policy on the graph. By our definitions, the deterministic policies* $\boldsymbol{f}_2 : \{1,2,3,4\} \to \{2,1,4,3\}$ *and* $\boldsymbol{f}_3 : \{1,2,3,4\} \to \{2,3,4,3\}$ *are a short cycle and a noose cycle, respectively. We define the randomised policy* $\boldsymbol{f}_4 = 1/2\boldsymbol{f}_2 + 1/2\boldsymbol{f}_3$, *which takes deterministic actions at vertices 1, 3 and 4, and randomised actions at vertex 2. The probability transition matrices associated with these policies are*

$$\mathbf{P}(\boldsymbol{f}_{\mathrm{H}}) = \begin{bmatrix} \cdot & 1 & \cdot & \cdot \\ \cdot & \cdot & 1 & \cdot \\ \cdot & \cdot & \cdot & 1 \\ 1 & \cdot & \cdot & \cdot \end{bmatrix}, \quad \mathbf{P}(\boldsymbol{f}_2) = \begin{bmatrix} \cdot & 1 & \cdot & \cdot \\ 1 & \cdot & \cdot & \cdot \\ \cdot & \cdot & \cdot & 1 \\ \cdot & \cdot & 1 & \cdot \end{bmatrix},$$

$$\mathbf{P}(\boldsymbol{f}_3) = \begin{bmatrix} \cdot & 1 & \cdot & \cdot \\ \cdot & \cdot & 1 & \cdot \\ \cdot & \cdot & \cdot & 1 \\ \cdot & \cdot & 1 & \cdot \end{bmatrix}, \quad \mathbf{P}(\boldsymbol{f}_4) = \begin{bmatrix} \cdot & 1 & \cdot & \cdot \\ 1/2 & \cdot & 1/2 & \cdot \\ \cdot & \cdot & \cdot & 1 \\ \cdot & \cdot & 1 & \cdot \end{bmatrix},$$

By (4.11), for $\beta = 0.9$ *the discounted rewards generated by* $\boldsymbol{f}_{\mathrm{H}}, \boldsymbol{f}_2, \boldsymbol{f}_3$ *and* $\boldsymbol{f}_4$, *respectively, are*

$$\begin{aligned} v_\beta(\boldsymbol{e}_1^{\mathrm{T}}, \boldsymbol{f}_{\mathrm{H}}) &= \{\boldsymbol{e}_1^{\mathrm{T}}[\mathbf{I} - \beta\mathbf{P}(\boldsymbol{f}_{\mathrm{H}})]^{-1}\}_1 \approx 2.91, \\ v_\beta(\boldsymbol{e}_1^{\mathrm{T}}, \boldsymbol{f}_2) &= \{\boldsymbol{e}_1^{\mathrm{T}}[\mathbf{I} - \beta\mathbf{P}(\boldsymbol{f}_2)]^{-1}\}_1 \approx 5.26, \\ v_\beta(\boldsymbol{e}_1^{\mathrm{T}}, \boldsymbol{f}_3) &= \{\boldsymbol{e}_1^{\mathrm{T}}[\mathbf{I} - \beta\mathbf{P}(\boldsymbol{f}_3)]^{-1}\}_1 \approx 1, \\ v_\beta(\boldsymbol{e}_1^{\mathrm{T}}, \boldsymbol{f}_4) &= \{\boldsymbol{e}_1^{\mathrm{T}}[\mathbf{I} - \beta\mathbf{P}(\boldsymbol{f}_4)]^{-1}\}_1 \approx 1.68. \end{aligned}$$

These values indicate that the usual objective function of maximising (or minimising) the reward (or cost) in a discounted MDP need not be optimal at Hamiltonian cycles.

4.3 Occupational Measures

In this section, we derive some characteristic properties of Hamiltonian policies. These arise in the context of the discounted MDP into which a given graph G has been embedded as in Section 4.2. The *occupation measure space*, $\mathcal{X}_\beta = \{\boldsymbol{x}(\boldsymbol{f}) | \boldsymbol{f} \in \mathcal{F}_S\}$ (induced by stationary policies) consists of vectors $\boldsymbol{x}(\boldsymbol{f})$ entries of which are *occupation measures of the state-action pairs* $(i, a) \in \mathcal{A}(i)$ defined by

$$x_{ia}(\boldsymbol{f}) = \{\boldsymbol{\nu}[(\mathbf{I} - \beta \mathbf{P}(\boldsymbol{f}))^{-1}]\}_i f_{ia}, \tag{4.13}$$

where $\boldsymbol{\nu} = (\nu_1, \ldots, \nu_N)$ denotes an arbitrary initial state distribution.

In what follows, we consider a specially structured initial distribution. Namely, for $\mu \in (0, 1/N)$ we define

$$\nu_i = \begin{cases} 1 - (N-1)\mu & \text{if } i = 1, \\ \mu & \text{otherwise.} \end{cases} \tag{4.14}$$

The *occupation measure of the state* i is defined as the aggregate

$$x_i(\boldsymbol{f}) = \sum_{a \in \mathcal{A}(i)} x_{ia}(\boldsymbol{f}) = \left\{\boldsymbol{\nu}[\mathbf{I} - \beta \mathbf{P}(\boldsymbol{f})]^{-1}\right\}_i, \tag{4.15}$$

where the second last equality follows from the fact that $\sum_{a \in \mathcal{A}(i)} f_{ia} = 1$. In particular,

$$x_1(\boldsymbol{f}) = \sum_{a \in \mathcal{A}(1)} x_{1a}(\boldsymbol{f}) = \left\{\boldsymbol{\nu}[\mathbf{I} - \beta \mathbf{P}(f)]^{-1}\right\}_1 = v_\beta(\boldsymbol{\nu}, \boldsymbol{f}), \tag{4.16}$$

where $v_\beta(\boldsymbol{\nu}, \boldsymbol{f}) = \sum_i \nu(i) v_\beta(\boldsymbol{e}_i, \boldsymbol{f})$ and $\boldsymbol{r} = (1, 0, \ldots, 0)^{\mathrm{T}}$. Note that $x_{ia}(\boldsymbol{f}) = [v_\beta(\boldsymbol{\nu}, \boldsymbol{f})]_i f(i, a)$. The construction of $\boldsymbol{x}$ in (4.13) defines a map M of the policy space $\mathcal{F}_S$ into $\mathbb{R}^{|\mathcal{A}|}$ by

$$M(\boldsymbol{f}) = \boldsymbol{x}(\boldsymbol{f}).$$

It is well-known (see Hordijk and Kallenberg [65] and Filar and Vrieze [52]) that for $\boldsymbol{\nu} > \mathbf{0}$, the map M is invertible and its inverse M^{-1} is defined by

$$M^{-1}(\boldsymbol{x})(i, a) = f_{\boldsymbol{x}}(i, a) = x_{ia}/x_i. \tag{4.17}$$

It is also known that, in this case, extreme points of $\mathcal{X}_\beta$ are in one-to-one correspondence with deterministic policies of $\mathcal{F}_\mathcal{D}$. However, this important property is lost when entries of $\boldsymbol{\nu}$ are permitted to take on zero values.

Connection with Polytopes and Linear Programs

There is an important, nowadays standard, conversion of a discounted MDP to a linear program (e.g., see Hordijk and Kallenberg [65]). However, in the case of an MDP arising from our embedding of the given graph G, Proposition 4.1 suggests that Hamiltonian cycles can be sought among the extreme points of the following subset of the *occupation measure space* $\mathcal{X}_\beta$ that is defined by the linear constraints

$$\sum_{i=1}^{N} \sum_{a \in \mathcal{A}(i)} \left(\delta_{ij} - \beta p(j|i,a)\right) x_{ia} = \nu_j, \quad j \in \mathcal{S}, \tag{4.18}$$

$$\sum_{a \in \mathcal{A}(1)} x_{1a} = \frac{(1-(N-1)\mu)(1-\beta) + \mu(\beta - \beta^N)}{(1-\beta)(1-\beta^N)}, \tag{4.19}$$

$$x_{ia} \geq 0, \quad i \in \mathcal{S}, a \in \mathcal{A}(i). \tag{4.20}$$

Proposition 4.1. *Let $\beta \in [0,1)$ and $\mu \in [0,1/N)$. Suppose that $\boldsymbol{f} \in \mathcal{F}_\mathcal{S}$ is Hamiltonian. The following two properties hold:*

(i) If the initial state distribution $\boldsymbol{\nu}$ is given by

$$\nu_i = \begin{cases} 1-(N-1)\mu & \text{if } i = 1, \\ \mu & \text{otherwise,} \end{cases} \tag{4.21}$$

then

$$v_\beta(\boldsymbol{\nu}, \boldsymbol{f}) = \frac{(1-(N-1)\mu)(1-\beta) + \mu(\beta - \beta^N)}{(1-\beta)(1-\beta^N)}. \tag{4.22}$$

(ii) If $\boldsymbol{f} \in \mathcal{F}_\mathcal{D}$ and $v_\beta(\boldsymbol{\nu}, \boldsymbol{f})$ is given by (4.22), then $\boldsymbol{f}$ is a Hamiltonian policy.

Proof. For $\mu = 0$, Part (i) is established in Feinberg [44]. For $\mu \in (0, 1/N)$, this part is merely an extension of the case when $\mu = 0$. Part (ii) follows by the same argument for the analogous result in [44]. □

4.4 Extreme Points and 1-randomised Policies

In this section, we examine policies induced by extreme points of the feasible polytope of $\mathcal{X}_\beta$. Our main aim of this section is to demonstrate that these policies are either Hamiltonian cycles or *1-randomised policies* satisfying a certain special property. A policy is said to be *1-randomised* if it is deterministic at all vertices except one $i \in \mathcal{S}$, and that vertex i is called a *splitting vertex* [46].

To ensure that the map M is always invertible and still distinguishes the home vertex 1 from all other vertices in $\mathcal{S}$, we define components of the initial distribution $\boldsymbol{\nu}^{(0)}$ by (4.14), for $\mu \in (0, 1/(N-1))$. Thus, we have a probability of $1-(N-1)\mu$ of starting at the home vertex 1 and of μ to start at each of the other vertices.

Remark 4.1. By choosing $\mu \in (0, 1/(N-1))$, we ensure that $\boldsymbol{\nu}^0$ is a proper distribution vector, where $\nu_i^0 \in (0,1)$ for all $i \in \mathcal{S}$. For $\mu \in (0, 1/N)$ (or $\mu \in (1/N, 1/(N-1)))$, we have a higher (or lower) probability of starting at the home vertex than at each of other vertices, respectively. Only when $\mu = 1/N$ do we have an equal probability of starting from anywhere.

Proposition 4.2. *Consider a Hamiltonian graph G of order N, $\beta \in (0,1)$, and $\boldsymbol{r} = (1, 0, \ldots, 0)^{\mathrm{T}}$. Then, for every $\mu \in (0, \mu_0) \subseteq (0, 1/(N-1))$ and with an initial distribution vector $\boldsymbol{\nu}^{(0)}$ defined as in* (4.14)*, the following two statements hold:*

(i) For any Hamiltonian policy $\boldsymbol{f}_{\mathrm{H}} \in \mathcal{F}_{\mathcal{S}}$,

$$v_\beta(\boldsymbol{\nu}^{(0)}, \boldsymbol{f}_{\mathrm{H}}) = \frac{(1-(N-1)\mu)(1-\beta) + \mu(\beta - \beta^N)}{(1-\beta)(1-\beta^N)}. \tag{4.23}$$

(ii) If a deterministic policy $\boldsymbol{f} \in \mathcal{F}_{\mathcal{D}}$ satisfies (4.23)*, then it is Hamiltonian.*

Proof. (i) Without loss of generality, let $\boldsymbol{f}_{\mathrm{H}} \in \mathcal{F}_{\mathcal{S}}$ trace out the standard Hamiltonian cycle $1 \to 2 \to \cdots \to N \to 1$. Recall that with the constant reward function $\boldsymbol{r} = (1, 0, \ldots, 0)^{\mathrm{T}}$, in a discounted MDP we receive a contribution of β^n for being at the home vertex at the nth step and 0 otherwise. If we start at the home vertex, we will return to it at every (nN)th step, for $n \in \mathbb{N}$. If we start at any other vertex $i \neq 1 \in \mathcal{S}$, we will return to the home vertex at $(nN-(i-1))$th step, for $n \in \mathbb{N}$. Therefore,

$$\begin{aligned}
&v_\beta(\boldsymbol{\nu}^{(0)}, \boldsymbol{f}_{\mathrm{H}}) \\
&= \{1-(N-1)\mu\}\left(1+\beta^N+\beta^{2N}+\cdots\right) + \mu\left(\beta^{N-1}+\beta^{2N-1}+\beta^{3N-1}+\cdots\right) \\
&+ \mu\left(\beta^{N-2}+\beta^{2N-2}+\beta^{3N-2}+\cdots\right) + \cdots + \mu\left(\beta+\beta^{N+1}+\beta^{2N+1}+\cdots\right) \\
&= \frac{\{1-(N-1)\mu\}(1-\beta)+\mu(\beta-\beta^N)}{(1-\beta)(1-\beta^N)}.
\end{aligned}$$

(ii) If $\boldsymbol{f}$ does not correspond to a Hamiltonian cycle, then it corresponds to either a noose cycle or a short cycle. A calculation similar to that in the proof of Part (i) shows that for a noose cycle,

$$v_\beta(\boldsymbol{\nu}^{(0)}, \boldsymbol{f}) = 1 - (N-1)\mu, \tag{4.24}$$

which is strictly less than $v_\beta(\boldsymbol{\nu}^{(0)}, \boldsymbol{f}_{\mathrm{H}})$. For a short cycle of length $k < N$, one similarly obtains

$$v_\beta(\boldsymbol{\nu}^{(0)}, \boldsymbol{f}) = \frac{\{1-(N-1)\mu\}(1-\beta) + \mu(\beta - \beta^k)}{(1-\beta)(1-\beta^k)}, \tag{4.25}$$

which is strictly larger than $v_\beta(\boldsymbol{\nu}^{(0)}, \boldsymbol{f}_{\mathrm{H}})$. The somewhat tedious details of the derivation of (4.24) and (4.25) can be found in Nguyen [81]. This proves the claim. □

By Proposition 4.2, we can define a subset $\widetilde{\mathcal{X}}_\beta$ of the discounted frequency space $\mathcal{X}_\beta$ characterised as the set of feasible solutions of the following system of linear constraints:

$$\sum_{i=1}^{N} \sum_{a \in \mathcal{A}(i)} (\delta_{ij} - \beta p(j|i,a)) x_{ia} = \nu_i, \text{ for all } i \in \mathcal{S}, \tag{4.26}$$

$$\sum_{a \in \mathcal{A}(1)} x_{1a} = \frac{(1-(N-1)\mu)(1-\beta) + \mu(\beta - \beta^N)}{(1-\beta)(1-\beta^N)}, \tag{4.27}$$

$$x_{ia} \geq 0, \text{ for all } i \in \mathcal{S}, a \in \mathcal{A}(i). \tag{4.28}$$

Formally,

$$\widetilde{\mathcal{X}}_\beta = \{\boldsymbol{x} | \boldsymbol{x} \text{ satisfies constraints (4.26)–(4.28)}\}, \tag{4.29}$$

the extreme points of which include Hamiltonian cycles on G. It is well-known (see Ross [93]) that every extreme point $\boldsymbol{x} \in \mathcal{X}_\beta$ has exactly N positive entries and $x_{ia} > 0$ for only one value of a for every $i \in \mathcal{S}$. This implies that using the inverse map M^{-1}, $\boldsymbol{f}(\boldsymbol{x})$ is a deterministic policy. As we introduce the additional constraint (4.27), an extreme point $\boldsymbol{x}$ of $\widetilde{\mathcal{X}}_\beta$ has N or $N+1$ positive entries; if the latter is true, then $x_{ia} > 0$ for two values of a for some $i \in \mathcal{S}$.

Consequently, using M^{-1}, $\boldsymbol{f}(\boldsymbol{x})$ is either a deterministic policy and therefore, by Proposition 4.2, is Hamiltonian, or it is a 1-randomised policy. Naturally, as these 1-randomised policies prevent us from employing a feasibility of $\widetilde{\mathcal{X}}_\beta$ to solve the HCP, it is important to understand their structures. Let $\boldsymbol{f}_\alpha$ be a 1-randomised policy induced by an extreme point $\boldsymbol{x}$ of $\widetilde{\mathcal{X}}_\beta$, where randomisation occurs at exactly one splitting vertex $i \in \mathcal{S}$, over two actions that are chosen with probability α and $1-\alpha$ for $\alpha \in (0,1)$. Such a 1-randomised policy can always be decomposed into two deterministic policies $\boldsymbol{f}_1$ and $\boldsymbol{f}_2$ such that

$$\boldsymbol{f}_\alpha = \alpha \boldsymbol{f}_1 + (1-\alpha)\boldsymbol{f}_2. \tag{4.30}$$

Note that $\boldsymbol{f}_1$ and $\boldsymbol{f}_2$ share the same actions at all vertices except the splitting vertex.

Remark 4.2. This raises an interesting, still open, question about the cardinality of the set of extreme points of $\widetilde{\mathcal{X}}_\beta$ induced by 1-randomised policies.

Theorem 4.1. *Consider a Hamiltonian graph G of order N, $\beta \in (0,1)$, and $\boldsymbol{r} = (1,0,\ldots,0)^{\mathrm{T}}$. For every $\mu \in (0,\mu_0) \subseteq (0,1/(N-1))$ and with an initial distribution vector $\boldsymbol{\nu}^{(0)}$ defined as in (4.14), let an extreme point $\boldsymbol{x}$ of $\widetilde{\mathcal{X}}_\beta$ induce a 1-randomised policy $\boldsymbol{f}_\alpha$ via the inverse map M^{-1}, $\alpha \in (0,1)$, and $\boldsymbol{f}_1$ and $\boldsymbol{f}_2$ be two deterministic policies satisfying $\boldsymbol{f}_\alpha = \alpha\boldsymbol{f}_1 + (1-\alpha)\boldsymbol{f}_2$. Then, the policies $\boldsymbol{f}_1$ and $\boldsymbol{f}_2$ cannot both be short cycles or noose cycles.*

Proof. The proof of this theorem follows immediately from that of Part (ii) of Proposition (4.2). □

Theorem 4.2. *Let $\boldsymbol{x}$ be an extreme point of $\widetilde{\mathcal{X}}_\beta$. Then, there exists a unique $\alpha \in (0,1)$ such that $\boldsymbol{x} = \boldsymbol{x}(\boldsymbol{f}_\alpha)$ where $\boldsymbol{f}_\alpha = \alpha\boldsymbol{f}_1 + (1-\alpha)\boldsymbol{f}_2$, with the policy $\boldsymbol{f}_1$ corresponding to a short cycle, the policy $\boldsymbol{f}_2$ corresponding to a noose cycle, and $\boldsymbol{f}_1$ and $\boldsymbol{f}_2$ coinciding at all vertices except at the splitting vertex i. In particular, α is given by (4.33), below.*

Proof. Let $\boldsymbol{x} = \{x_{ia}\}$ and $\boldsymbol{y} = \{y_{ia}\}$ be two discounted occupation measures corresponding to the discount factor β and initial distribution vector $\boldsymbol{\nu}^{(0)}$. We define $x_i = \sum_{a \in \mathcal{A}(i)} x_{ia}$ and $y_i = \sum_{a\in\mathcal{A}(i)} y_{ia}$, for $i \in \mathcal{S}$. Applying the inverse map defined as in (4.17), we have

$$f_{\boldsymbol{x}}(i,a) = x_{ia}/x_i \quad \text{and} \quad f_{\boldsymbol{y}}(i,a) = y_{ia}/y_i.$$

Let $\gamma \in [0,1]$ and define $\boldsymbol{z} = \gamma\boldsymbol{x} + (1-\gamma)\boldsymbol{y}$. Then, since

$$x_i = \nu_i^{(0)} + \beta \sum_j x_{ja} p(i|j,a),$$
$$y_i = \nu_i^{(0)} + \beta \sum_j y_{ja} p(i|j,a),$$

we also have

$$z_i = \nu_i^{(0)} + \beta \sum_j z_{ja} p(i|j,a),$$

that is, $\boldsymbol{z}$ is a discounted occupation measure as well. If $\boldsymbol{x}$ and $\boldsymbol{y}$ correspond to deterministic policies such that $\boldsymbol{z}$ is an optimal randomisation, that is, $\boldsymbol{z}$

is an extreme point of $\widetilde{\mathcal{X}}_\beta$, then we have an explicit expression for

$$z_1 = \gamma x_1 + (1-\gamma) y_1. \tag{4.31}$$

given by (4.16) and (4.23). On the other hand, since one of $\boldsymbol{x}, \boldsymbol{y}$—say, $\boldsymbol{x}$—corresponds to a short cycle, we have an explicit expression for x_1, given by (4.25). The same is true for y_1, the expression of which is given by (4.24) as $\boldsymbol{y}$ corresponds to a noose cycle. Hence, (4.31) gives an explicit expression for γ. Furthermore, by defining $z_i = \sum_{a \in \mathcal{A}(i)} z_{ia}$ and considering the (i,a)th entry of $\boldsymbol{z}$ divided by z_{ia}, we obtain

$$\begin{aligned} f_{\boldsymbol{z}}(i,a) &= z_{ia}/z_i \\ &= \gamma x_{ia}/z_i + (1-\gamma) y_{ia}/z_i \\ &= \{\frac{\gamma x_i}{\gamma x_i + (1-\gamma) y_i}\} f_{\boldsymbol{x}}(i,a) + \{\frac{(1-\gamma) y_i}{\gamma x_i + (1-\gamma) y_i}\} f_{\boldsymbol{y}}(i,a), \end{aligned} \tag{4.32}$$

as $\boldsymbol{z} = \gamma \boldsymbol{x} + (1-\gamma)\boldsymbol{y}$. From (4.32),

$$\alpha = \frac{\gamma x_i}{\gamma x_i + (1-\gamma) y_i}, \tag{4.33}$$

which can thus be explicitly calculated since γ is already known. □

4.5 A Parameter-Free Model

Corresponding to each given graph G, we can construct a discounted MDP. It is well known that each discounted MDP is associated with the following system of linear equations

$$\boldsymbol{x_f}(\beta)^{\mathrm{T}}\{\mathbf{I} - \beta \mathbf{P}(\boldsymbol{f})\} = \boldsymbol{\nu}, \tag{4.34}$$

where $\boldsymbol{x_f}(\beta)$ is a vector with components $x_i(\boldsymbol{f})$ defined by (4.15), and $\boldsymbol{\nu}$ is the initial distribution. Equivalently,

$$\sum_{i=1}^{N} x_i(\delta_{ij} - \beta p_{ij}) = \nu_j, \qquad \text{for } j \in \mathcal{S}. \tag{4.35}$$

Furthermore, by assuming $\boldsymbol{x_f}(\beta) > \mathbf{0}$, component-wise, and defining $x_{ia} = x_i f(i,a)$, we can rewrite the left hand side of (4.35) as below

$$\sum_{i=1}^{N} x_i(\delta_{ij} - \beta p_{ij}) = \sum_{i=1}^{N} x_i \sum_{a \in \mathcal{A}(i)} (\delta_{ij} - \beta p(j|i,a)) f(i,a)$$

$$
\begin{aligned}
&= \sum_{i=1}^{N} x_i \sum_{a \in \mathcal{A}(i)} (\delta_{ij} - \beta p(j|i,a)) x_{ia}/x_i \\
&= \sum_{i=1}^{N} \sum_{a \in \mathcal{A}(i)} (\delta_{ij} - \beta p(j|i,a) x_{ia} \\
&= \sum_{a \in \mathcal{A}(j)} x_{ja} - \beta \sum_{b \in \mathcal{B}(j)} x_{bj},
\end{aligned}
$$

where the last equality is derived from (4.12). Thus, (4.35) is simplified to

$$\sum_{a \in \mathcal{A}(j)} x_{ja} - \beta \sum_{b \in \mathcal{B}(j)} x_{bj} = \nu_j, \qquad \text{for } j \in \mathcal{S}. \tag{4.36}$$

Next, we recognise that constraints (4.36) could be analogously derived for N distinct, degenerate initial distributions $\boldsymbol{\nu} = \boldsymbol{e}_k^{\mathrm{T}}$, $k = 1, 2, \ldots, N$, where $\boldsymbol{e}_k$ is a column vector with 1 in the kth entry and 0 elsewhere. Furthermore, it will be convenient to multiply all the constraints in (4.36) by a normalising factor of $1 - \beta^N$ and make a change of variables by setting

$$x_{ja}^k = (1-\beta) x_{ja}, \qquad \text{for } a \in \mathcal{A}(j), j \in \mathcal{S},$$

where k indicates that the initial state distribution is $\boldsymbol{e}_k^{\mathrm{T}}$. Hence, constraints (4.36) now take the form

$$\sum_{a \in \mathcal{A}(j)} x_{ja}^k - \beta \sum_{b \in \mathcal{B}(j)} x_{bj}^k = (1-\beta)^N \delta_{jk}, \quad \text{for } j \in \mathcal{S}, \tag{4.37}$$

where δ_{kj} is the Kronecker delta. It should be noted that there now might be feasible non-negative solutions of (4.37) such that

$$x_j^k = \sum_{a \in \mathcal{A}(j)} x_{ja}^k = 0,$$

for some, but not all, $j \in \mathcal{S}$. Nonetheless, from each non-negative solution of (4.37) we may construct a vector $\boldsymbol{x}^k(\beta) = (x_1^k, \ldots, x_N^k)$, where $x_j^k = \sum_{a \in \mathcal{A}(j)} x_{ja}^k$. Then, it is still possible to associate a stationary policy $\boldsymbol{f}_{\boldsymbol{x}^k}$ defined, in the spirit of (4.17), by

$$f_{\boldsymbol{x}^k}(j,a) = \begin{cases} x_{ja}^k / x_j^k & \text{if } x_j^k > 0, \\ \text{arbitrary} & \text{otherwise.} \end{cases}$$

Now, the policy $\boldsymbol{f}_{\boldsymbol{x}^k}$ still induces a Markov chain with the probability transition matrix $\mathbf{P}(\boldsymbol{f}_{\boldsymbol{x}^k})$ (see (4.4)). It is straightforward to verify that in matrix notation, (4.37) become

$$[\boldsymbol{x}^k(\beta)]^{\mathrm{T}}(\mathbf{I}-\beta\mathbf{P}) = (1-\beta^N)\boldsymbol{e}_k^{\mathrm{T}}, \quad k \in \mathcal{S}. \tag{4.38}$$

If $\boldsymbol{x}^k$ has exactly N positive entries that are in one-to-one correspondence with a Hamiltonian cycle, then $\mathbf{P}$ is Hamiltonian (an irreducible permutation matrix), $\mathbf{P}^N = \mathbf{I}$ the identity matrix and it follows that

$$(\mathbf{I}-\beta\mathbf{P})^{-1} = \frac{1}{1-\beta^N}\{\mathbf{I}+\beta\mathbf{P}+\cdots+\beta^{N-1}\mathbf{P}^{N-1}\}. \tag{4.39}$$

Consequently, (4.38) and (4.39) imply that

$$\begin{aligned}\boldsymbol{x}^k(\beta)^{\mathrm{T}} &= (1-\beta^N)\boldsymbol{e}_k^{\mathrm{T}}(\mathbf{I}-\beta\mathbf{P})^{-1}\\ &= \boldsymbol{e}_k^{\mathrm{T}}\mathbf{I}+\beta\boldsymbol{e}_k^{\mathrm{T}}\mathbf{P}+\beta^2\boldsymbol{e}_k^{\mathrm{T}}\mathbf{P}^2+\cdots+\beta^{N-1}\boldsymbol{e}_k^{\mathrm{T}}\mathbf{P}^{N-1}.\end{aligned} \tag{4.40}$$

Define a new vector $\boldsymbol{x}_r^k = \boldsymbol{e}_k^{\mathrm{T}}\mathbf{P}^r$ for $r = 0, 1, \ldots, N-1$. Clearly, all components of vector $\boldsymbol{x}_r^k$ are equal to 0 except for one that is equal to 1. This unique element identifies the rth vertex visited on a Hamiltonian cycle starting from vertex k. Hence, we can rewrite $\boldsymbol{x}^k(\beta)$ in terms of vectors $\boldsymbol{x}_r^k$ as follows

$$\boldsymbol{x}^k(\beta) = \boldsymbol{x}_0^k+\beta\boldsymbol{x}_1^k+\beta^2\boldsymbol{x}_2^k+\cdots+\beta^{N-1}\boldsymbol{x}_{N-1}^k. \tag{4.41}$$

Replacing $\boldsymbol{x}^k(\beta)$ with the right-hand side of (4.41) in (4.38), we obtain

$$(\boldsymbol{x}_0^k+\beta\boldsymbol{x}_1^k+\beta^2\boldsymbol{x}_2^k+\cdots+\beta^{N-1}\boldsymbol{x}_{N-1}^k)^{\mathrm{T}}(\mathbf{I}-\beta\mathbf{P}) = (1-\beta^N)\boldsymbol{e}_k^{\mathrm{T}}$$

for $k \in \mathcal{S}$. Equivalently,

$$\begin{aligned}&\boldsymbol{x}_0^k+\beta(\boldsymbol{x}_1^k-\mathbf{P}^{\mathrm{T}}\boldsymbol{x}_0^k)+\beta^2(\boldsymbol{x}_2^k-\mathbf{P}^{\mathrm{T}}\boldsymbol{x}_1^k)+\cdots\\ &\quad+\beta^{N-1}(\boldsymbol{x}_{N-1}^k-\mathbf{P}^{\mathrm{T}}\boldsymbol{x}_{N-2}^k)-\beta^N\mathbf{P}^{\mathrm{T}}\boldsymbol{x}_{N-1}^k = \boldsymbol{e}_k-\beta^N\boldsymbol{e}_k.\end{aligned} \tag{4.42}$$

Now, by equating coefficients of powers of β in both sides of the latter equation, we obtain the following set of linear constraints that are free of the parameter β:

$$\boldsymbol{x}_0^k = \boldsymbol{e}_k, \tag{4.43}$$

$$\boldsymbol{x}_r^k-\mathbf{P}^{\mathrm{T}}\boldsymbol{x}_{r-1}^k = \mathbf{0}, \quad \text{for } r \in \mathcal{S}\backslash\{1, N\}, \tag{4.44}$$

$$\mathbf{P}^{\mathrm{T}}\boldsymbol{x}_{N-1}^k = \boldsymbol{e}_k. \tag{4.45}$$

Analogous to the way that we derived (4.36), we can expand (4.43)–(4.45) as follows

$$\sum_{a\in\mathcal{A}(i)} x_{0,ia}^k = \delta_{ki}, \quad \text{for } i, k \in \mathcal{S}, \tag{4.46}$$

$$\sum_{c\in\mathcal{A}(i)} x_{r,ia}^k - \sum_{b\in\mathcal{B}(i)} x_{r-1,bi}^k = 0, \quad \text{for } i, k \in \mathcal{S}, r \in \mathcal{S}\backslash\{N\}, \tag{4.47}$$

$$\sum_{b\in\mathcal{B}(i)} x^k_{N-1,bi} = \delta_{ki}, \qquad \text{for } i,k \in \mathcal{S}. \tag{4.48}$$

Based on the way we defined vectors $\boldsymbol{x}^k_r$, we now have an interesting interpretation for components $x^k_{r,ia}$. In particular,

$$x^k_{r,ia} = \{\boldsymbol{e}^{\mathrm{T}}_k \mathbf{P}^r(\boldsymbol{f}_{\boldsymbol{x}^k})\boldsymbol{e}_i\} f_{\boldsymbol{x}^k}(i,a),$$

the probability that starting from k, state/vertex i is visited on the rth step and action $a \in \mathcal{A}(i)$ is selected. More precisely, if these vectors indeed came from a Hamiltonian transition matrix $\mathbf{P}$, then we may demand that $x^k_{r,ia}$ be equal to 1 if edge (i,a) is visited at the rth step of a Hamiltonian cycle starting from vertex k and otherwise equal to 0.

However, due to complexity associated with the introduction of 0–1 variables, we can relax them to vary between 0 and 1 and try to establish new linear constraints designed to force them to behave somewhat like 0–1 variables. New constraints which can be augmented by utilising this interpretation are listed below:

(i) If a vertex i is visited at rth step of a Hamiltonian cycle starting from vertex k, then on the same cycle, vertex k should be visited at $N-r$ steps starting from vertex i:

$$\sum_{a\in\mathcal{A}(i)} x^k_{r,ia} - \sum_{a\in\mathcal{A}(k)} x^i_{N-r,ka} = 0, \qquad \text{for } i,k \in \mathcal{S}, i \neq k, r \in \mathcal{S}\backslash\{N\}. \tag{4.49}$$

(ii) If an edge (i,a) belongs to a particular Hamiltonian cycle, then it should be visited from any starting vertex on that cycle, in particular from starting vertices k and j:

$$\sum_{r=0}^{N-1} x^k_{r,ia} - \sum_{r=0}^{N-1} x^j_{r,ia} = 0, \qquad \text{for } i,k,j \in \mathcal{S}, k < j, a \in \mathcal{A}(i). \tag{4.50}$$

Of course, if (i,a) does not belong to a Hamiltonian cycle, both terms on the left-hand side of (4.50) are 0.

(iii) If an edge (i,a) belongs to a particular Hamiltonian cycle, then based on the starting vertex, it can be visited at any steps of $0,1,\ldots,N-1$ on that cycle:

$$\sum_{k=1}^{N} x^k_{r,ia} - \sum_{k=1}^{N} x^k_{t,ia} = 0, \qquad \text{for } i \in \mathcal{S}, r,t = 0,1,\ldots,N-1, r < t, a \in \mathcal{A}(i). \tag{4.51}$$

Of course, if (i,a) does not belong to a Hamiltonian cycle, both terms on the left-hand side of (4.51) are zero.

(iv) Starting from vertex k, we must visit vertex i at some step in the next $N-1$ steps:

$$\sum_{r=0}^{N-1} \sum_{a \in \mathcal{A}(i)} x^k_{r,ia} = 1, \qquad \text{for } i, k \in \mathcal{S}. \tag{4.52}$$

(v) Starting from vertex k, we must visit exactly one vertex at rth step:

$$\sum_{i=1}^{N} \sum_{a \in \mathcal{A}(i)} x^k_{r,ia} = 1, \qquad \text{for } k \in \mathcal{S}, r = 0, 1, \ldots, N-1. \tag{4.53}$$

(vi) To tighten the whole feasible region, we can exploit the notion of a *shortest path*. For this purpose,we assign length 1 to all edges of graph G. Then, in polynomial time we can calculate the shortest path between each pair of vertices in this graph. If θ_{ki} is the length of the shortest path between vertices k and i in G, starting from vertex k, vertex i cannot be visited earlier than θ_{ki} steps. Moreover, starting from vertex k, vertex i cannot be visited later than $N - \theta_{ik}$ steps (on a Hamiltonian cycle) either:

$$x^k_{r,ia} = 0, \qquad \text{for } k, i \in \mathcal{S}, a \in \mathcal{A}(i), r < \theta_{ki}, r \geq N - \theta_{ak}. \tag{4.54}$$

(vii) If k is the starting vertex, it should not be revisited before the last step:

$$x^k_{r,ka} = 0, \qquad \text{for } k \in \mathcal{S}, a \in \mathcal{A}(k), r \in \mathcal{S} \backslash \{N\}, \tag{4.55}$$

$$x^k_{r,bk} = 0, \qquad \text{for } k \in \mathcal{S}, b \in \mathcal{B}(k), r = 0, 1, \ldots, N-2. \tag{4.56}$$

(viii) Finally, all variables should be non-negative:

$$x^k_{r,ia} \geq 0, \qquad \text{for } i, k \in \mathcal{S}, r = 0, 1, \ldots, N-1, a \in \mathcal{A}(i). \tag{4.57}$$

We denote by $\mathcal{PF}$ the polytope defined by the set of linear constraints (4.46)–(4.57). This model has $N^2|\mathcal{A}|$ variables and $2N^3 + N^2 + N + N(N-1)|\mathcal{A}|$ constraints (excluding non-negativity constraints). Thus, the numbers of variables and constraints are bounded above by $N^4 - N^3$ and $N^4 + 2N^2 + N$, respectively. Moreover, for the family of cubic graphs, the number of variables and constraints are $3N^3$ and $5N^3 - 2N^2 + N$, respectively.

Clearly, if we force all $x^k_{r,ia}$ variables to be binary variables, then the polytope $\mathcal{PF}$ will be infeasible for all non-Hamiltonian graphs. However, we hope that by relaxing such 0–1 condition, this model could still recognise non-Hamiltonian graphs by showing infeasibility. Next, we present numerical evidence that this goal is attained for a rather large class of non-Hamiltonian cubic graphs.

Numerical results

We have tested many Hamiltonian and non-Hamiltonian cubic graphs through polytope $\mathcal{PF}$. So far, it has been successful in recognising all bridge cubic graphs[2], on 10, 12, 14, 16 and 18 vertices by showing infeasibility. Surprisingly, it could also eliminate some of the non-bridge non-Hamiltonian (NBNH) cubic graphs by showing infeasibility. Table 4.1 displays the results of these experiments.

Table 4.1: Solving the HCP for non-Hamiltonian cubic graphs

N	Number of Bridge Graphs	Eliminated	Number of NBNH	Eliminated	Ratio
10	1	1	1	0	0
12	4	4	1	0	0
14	29	29	6	1	0.167
16	186	186	33	6	0.182
18	1435	1435	231	42	0.182

In Table 4.1, the first column indicates the number of vertices in the family of cubic graphs. Then, the second to fourth columns show the total number of bridge graphs in that particular family, the number of such bridge graphs that made the corresponding polytope $\mathcal{PF}$ infeasible, the total number of NBNH graphs in that particular family and the number of such NBNH graphs that made the corresponding polytope $\mathcal{PF}$ infeasible. The last column gives the ratio of the fifth column over the fourth one.

The promising results indicated in Table 4.1 strongly suggest the need for further work on improving the $\mathcal{PF}$ model.

[2] Recall that a *bridge* is an edge the removal of which disconnects the graph, and a *bridge graph* is a graph that contains one or more bridges.

Part III
Optimisation

Chapter 5
Determinants

5.1 Introduction

Ejov *et al.* [35] embed the Hamiltonian cycle problem in a limiting average Markov decision process model with an asymmetric linear perturbation defined in (3.14). The authors prove that for positive and sufficiently small values of ε, determining the Hamiltonicity of a given graph is equivalent to minimising the top-left element of the fundamental matrix $\mathbf{G}$ of an MDP associated with the given graph, over the space of feasible deterministic policies $\mathcal{F}_{\mathcal{D}}$. The conjecture that the same holds for the larger space $\mathcal{F}_{\mathcal{S}}$ remains open.

In Chapter 3, we replaced the asymmetric linear perturbation used in Ejov *et al.* [35] by a symmetric linear perturbation defined in (3.13), and extended the result to the set of feasible doubly stochastic policies $\mathcal{F}_{\mathcal{DS}}$. We also presented two mathematical programming problems, both of which are in the spirit of

$$\begin{array}{c}\min \mathbf{G}_{11} \\ \text{over the set of feasible doubly stochastic policies.}\end{array} \tag{5.1}$$

Theorem 3.6 shows that the gap between optimal values of (5.1) for a Hamiltonian graph and for a non-Hamiltonian graph of the same order is strictly positive (see also Borkar *et al.* [18]). This strengthens the potential usefulness of the optimisation model (5.1). However, in implementing an efficient algorithm for (5.1), numerical difficulties arise. The first major challenge is the evaluation of the $\mathbf{G}_{11}$ functional, as the fundamental matrix is an inverse. The second major challenge is that relevant Hessian matrices, an important ingredient in many optimisation algorithms, are dense due to perturbation and consequently expensive to compute.

Looking for other functionals to determine Hamiltonicity, we prove that the determinant of the inverse of the fundamental matrix, too, is an appropriate optimisation functional. Furthermore, we show that the determinant objective function is more robust than the function $\mathbf{G}_{11}$, as it is maximised at Hamiltonian cycles with or without the linear symmetric perturbation considered in Borkar *et al.* [17] and in Chapter 3.

5.2 Optimality at Hamiltonian Cycles

By (2.8) in Chapter 2, we define the matrix function $\mathbf{W}(\mathbf{P}) = \mathbf{I} - \mathbf{P} + 1/N\mathbf{J}$, where $\mathbf{J} = \boldsymbol{e}\boldsymbol{e}^{\mathrm{T}}$ is a matrix every element of which is 1. For a given policy $\boldsymbol{f}$, we define similarly the function

$$\mathbf{W}(\boldsymbol{f}) = \mathbf{I} - \mathbf{P}(\boldsymbol{f}) + 1/N\mathbf{J}. \tag{5.2}$$

We apply the symmetric linear perturbation defined in (3.13) to $\mathbf{P}(\boldsymbol{f})$ to obtain $\mathbf{P}^{\varepsilon}(\boldsymbol{f})$, that is,

$$\mathbf{P}^{\varepsilon}(\boldsymbol{f}) = (1-\varepsilon)\mathbf{P}(\boldsymbol{f}) + \varepsilon/N\mathbf{J}. \tag{5.3}$$

Recall from (4.6) that the fundamental matrix $\mathbf{G}(\boldsymbol{f})$ is $\{\mathbf{I} - \mathbf{P}(\boldsymbol{f}) + \mathbf{P}^*(\boldsymbol{f})\}^{-1}$, and from (4.5) that the stationary distribution matrix $\mathbf{P}^*(\boldsymbol{f})$ is

$$\mathbf{P}^*(\boldsymbol{f}) = \lim_{T\to\infty} \frac{1}{T+1} \sum_{t=0}^{T} \mathbf{P}^t(\boldsymbol{f}). \tag{5.4}$$

Then, the stationary distribution matrix of $\mathbf{P}^{\varepsilon}(\boldsymbol{f})$ is

$$\mathbf{P}^*(\boldsymbol{f},\varepsilon) = \lim_{T\to\infty} \frac{1}{T+1} \sum_{t=0}^{T} (\mathbf{P}^{\varepsilon})^t(\boldsymbol{f}), \tag{5.5}$$

the fundamental matrix of $\mathbf{P}^{\varepsilon}(\boldsymbol{f})$ is

$$\mathbf{G}(\boldsymbol{f},\varepsilon) = \{\mathbf{I} - \mathbf{P}^{\varepsilon}(\boldsymbol{f}) + \mathbf{P}^*(\boldsymbol{f},\varepsilon)\}^{-1}, \tag{5.6}$$

and,

$$\mathbf{W}(\boldsymbol{f},\varepsilon) = \mathbf{I} - \mathbf{P}^{\varepsilon}(\boldsymbol{f}) + 1/N\mathbf{J}. \tag{5.7}$$

We denote by $\mathbf{A}^{11}$ the matrix $\mathbf{A}$ with the first row and first column removed.

Lemma 5.1. *For any doubly stochastic policy* $\boldsymbol{f}$ *and* $\varepsilon > 0$,

$$\mathbf{G}_{11}(\boldsymbol{f},\varepsilon) = \frac{\det \mathbf{W}^{11}(\boldsymbol{f},\varepsilon)}{\det \mathbf{W}(\boldsymbol{f},\varepsilon)}. \tag{5.8}$$

Proof. For any doubly stochastic policy $\boldsymbol{f}$, $\mathbf{P}^{\varepsilon}(\boldsymbol{f})$ is doubly stochastic and consequently, $\mathbf{P}^*(\boldsymbol{f},\varepsilon) = 1/N\mathbf{J}$ by (3.9). Therefore,

$$\begin{aligned}\mathbf{G}(\boldsymbol{f},\varepsilon) &= \{\mathbf{I} - \mathbf{P}^{\varepsilon}(\boldsymbol{f}) + 1/N\mathbf{J}\}^{-1}\\ &= \mathbf{W}^{-1}(\boldsymbol{f},\varepsilon).\end{aligned} \tag{5.9}$$

We denote by p^{ε}_{ij} and g_{ij} the (i,j)th entries of $\mathbf{P}^*(\boldsymbol{f},\varepsilon)$ and $\mathbf{G}(\boldsymbol{f},\varepsilon)$, respectively. Obviously, these entries depend on $\boldsymbol{f}$, which we suppress for notational convenience. From (5.9), we have $\mathbf{W}(\boldsymbol{f},\varepsilon)\mathbf{G}(\boldsymbol{f},\varepsilon) = \mathbf{I}$, which leads to

$$\begin{aligned}\mathbf{W}(\boldsymbol{f},\varepsilon)&\begin{bmatrix} g_{11}\\ g_{21}\\ \vdots\\ g_{N1}\end{bmatrix}\\
&= \begin{bmatrix} (N+1)/N - p^{\varepsilon}_{11} & 1/N - p^{\varepsilon}_{12} & \cdots & 1/N - p^{\varepsilon}_{1N}\\ 1/N - p^{\varepsilon}_{21} & (N+1)/N - p^{\varepsilon}_{22} & \cdots & 1/N - p^{\varepsilon}_{2N}\\ \vdots & \vdots & \ddots & \vdots\\ 1/N - p^{\varepsilon}_{N1} & 1/N - p^{\varepsilon}_{N2} & \cdots & (N+1)/N - p^{\varepsilon}_{NN}\end{bmatrix}\begin{bmatrix} g_{11}\\ g_{21}\\ \vdots\\ g_{N1}\end{bmatrix}\\
&= \begin{bmatrix} 1\\ 0\\ \vdots\\ 0\end{bmatrix}.\end{aligned}$$

Applying Cramer's rule [66], we obtain

$$\begin{aligned}\mathbf{G}_{11}(\boldsymbol{f},\varepsilon) &= \frac{\det\left(\begin{bmatrix} 1 & 1/N - p^{\varepsilon}_{12} & \cdots & 1/N - p^{\varepsilon}_{1N}\\ 0 & (N+1)/N - p^{\varepsilon}_{22} & \cdots & 1/N - p^{\varepsilon}_{2N}\\ \vdots & \vdots & \ddots & \vdots\\ 0 & 1/N - p^{\varepsilon}_{N2} & \cdots & (N+1)/N - p^{\varepsilon}_{NN}\end{bmatrix}\right)}{\det \mathbf{W}(\boldsymbol{f},\varepsilon)}\\
&= \frac{\det \mathbf{W}^{11}(\boldsymbol{f},\varepsilon)}{\det \mathbf{W}(\boldsymbol{f},\varepsilon)}.\end{aligned}$$

□

Lemma 5.1 makes it plausible that, perhaps, instead of minimising $\mathbf{G}_{11}(\boldsymbol{f},\varepsilon)$ over $\mathcal{F}_{\mathcal{DS}}$, we can also maximise $\det \mathbf{W}(\boldsymbol{f},\varepsilon)$ over the same set to determine if a given graph is Hamiltonian. This conjecture turns out to be correct for $\varepsilon \in [0,1)$ over $\mathcal{F}_{\mathcal{S}}$. We generalise this result further by replacing the coefficient $1/N$ with $\alpha > 0$, to obtain the matrix

$$\mathbf{W}_{\alpha}(\boldsymbol{f},\varepsilon) = \mathbf{I} - \mathbf{P}^{\varepsilon}(\boldsymbol{f}) + \alpha\mathbf{J}. \tag{5.10}$$

the determinant value of which, too, is maximised at Hamiltonian cycles. Recall that the circumference of a graph is the length of a longest cycle in the graph. The following theorem shows the relationship between the matrix function $\det \mathbf{W}_{\alpha}(\boldsymbol{f},\varepsilon)$ and circumferences of graphs.

Theorem 5.1. *For any feasible policy $\boldsymbol{f}$ on a given graph of order N,*

$$\det \mathbf{W}_{\alpha}(\boldsymbol{f},\varepsilon) \leq \alpha N \frac{1-(1-\varepsilon)^k}{\varepsilon}, \tag{5.11}$$

where $k \leq N$ is the circumference of the graph and $\varepsilon \in [0,1)$.

For clarity, here we consider only the case $\alpha = 1/N$. For the general proof of the case $\alpha \in \mathbb{R}^+$, we refer the interested reader to Ejov *et al.* [34] and Nguyen [81]. We prove Theorem 5.1 for the cases of $\varepsilon = 0$ and $\varepsilon > 0$ in Sections 5.2.1 and 5.2.2, respectively.

Consider the following decomposition of the set $\mathcal{F}_{\mathcal{D}}$ of deterministic policies: $\mathcal{F}_{\mathcal{D}} = \{\mathcal{F}_{\mathcal{H}} \cup \mathcal{F}_{\mathcal{D}_1} \cup \mathcal{F}_{\mathcal{D}_2}\}$, where $\mathcal{F}_{\mathcal{H}}, \mathcal{F}_{\mathcal{D}_1}$, and $\mathcal{F}_{\mathcal{D}_2}$ are three mutually exclusive (possibly empty) subsets of $\mathcal{F}_{\mathcal{D}}$ that contain Hamiltonian cycles, policies possessing a single cycle of length $k < N$, and policies possessing two or more disjoint cycles, respectively. Figure 5.1 illustrates this decomposition of the set $\mathcal{F}_{\mathcal{S}}$.

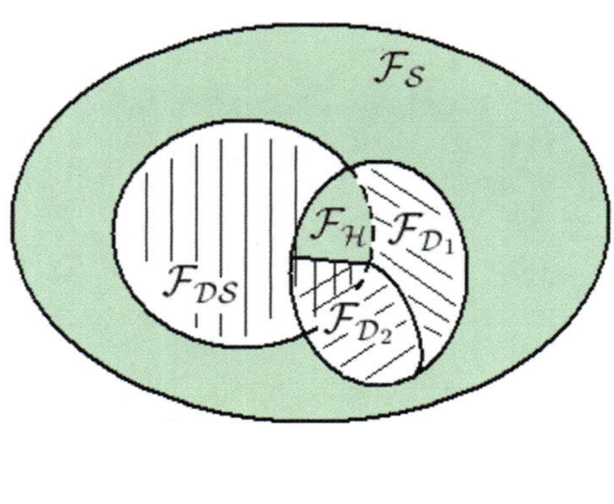

Fig. 5.1: A Venn diagram of $\mathcal{F}_{\mathcal{S}}$

5.2.1 Unperturbed Case

We restate Theorem 5.1 for $\varepsilon = 0$ and $\alpha = 1/N$.

Theorem 5.2. *For any feasible policy $\boldsymbol{f}$ of a given graph of order N,*

$$\det \mathbf{W}(\boldsymbol{f}) \leq k, \tag{5.12}$$

where $k \leq N$ is the circumference of the graph.

The proof of Theorem 5.2 requires the following intermediate results: Corollary 5.1 states a relationship between eigenvalues of $\mathbf{W}(\boldsymbol{f})$ and those of $\mathbf{P}(\boldsymbol{f})$; Propositions 5.1–5.3 establish that inequality (5.12) holds for a deterministic policy $\boldsymbol{f} \in \mathcal{F}_{\mathcal{H}}$, $\mathcal{F}_{\mathcal{D}_1}$ and $\mathcal{F}_{\mathcal{D}_2}$, respectively. Collectively, these results imply that the upper bound (5.12) is true for any $\boldsymbol{f} \in \mathcal{F}_{\mathcal{D}}$; Proposition 5.4 shows that given a randomised policy $\boldsymbol{f} \in \mathcal{F}_{\mathcal{S}} \setminus \mathcal{F}_{\mathcal{D}}$, there exist two deterministic policies $\bar{\boldsymbol{f}}$ and $\hat{\boldsymbol{f}} \in \mathcal{F}_{\mathcal{D}}$ such that

$$\det \mathbf{W}(\bar{\boldsymbol{f}}) \leq \det \mathbf{W}(\boldsymbol{f}) \leq \det \mathbf{W}(\hat{\boldsymbol{f}}). \tag{5.13}$$

Inequalities (5.13), together with Propositions 5.1–5.3, indicate that the upper bound (5.12) also holds for any given randomised policy $\boldsymbol{f} \in \mathcal{F}_{\mathcal{S}} \setminus \mathcal{F}_{\mathcal{D}}$.

For $\boldsymbol{f} \in \mathcal{F}_{\mathcal{S}}$ and $i = 1, \ldots, N$, let λ_i's and μ_i's be the eigenvalues of $\mathbf{P}(\boldsymbol{f})$ and $\mathbf{W}(\boldsymbol{f})$, respectively. As $\mathbf{P}(\boldsymbol{f})$ is stochastic, at least one eigenvalue of $\mathbf{P}(\boldsymbol{f})$ is unity, and we denote this eigenvalue by λ_N.

Corollary 5.1. *For any feasible policy $\boldsymbol{f} \in \mathcal{F}_{\mathcal{S}}$ on a given graph of order N, eigenvalues μ_i's of $\mathbf{W}(\boldsymbol{f})$ are*

$$\mu_i = \begin{cases} 1 - \lambda_i & \text{for } i = 1, \ldots, N-1, \\ 0 & \text{for } i = N. \end{cases} \tag{5.14}$$

Proof. This result follows from Serra-Capizzano [96, Theorem 2.3]. □

Corollary 5.2. *For any feasible policy $\boldsymbol{f} \in \mathcal{F}_{\mathcal{S}}$ on a given graph of order N,*

$$\det \mathbf{W}(\boldsymbol{f}) = \prod_{i=1}^{N-1} (1 - \lambda_i). \tag{5.15}$$

Proof. It is well-known that

$$\det \mathbf{W}(\boldsymbol{f}) = \prod_{i=1}^{N} \mu_i. \tag{5.16}$$

Then, Corollary 5.2 follows immediately from Corollary 5.1. □

Proposition 5.1. *For any Hamiltonian policy $\boldsymbol{f}_{\mathrm{H}} \in \mathcal{F}_{\mathcal{H}}$ on a given graph of order N,*

$$\det \mathbf{W}(\boldsymbol{f}_{\mathrm{H}}) = N. \tag{5.17}$$

Proof. By Corollary 5.2,

$$\begin{aligned}\det \mathbf{W}(\boldsymbol{f}_{\mathrm{H}}) &= \prod_{i=1}^{N-1}(1-\lambda_i)\\ &= 1 - \sum_{i=1}^{N-1}\lambda_i + \sum_{\substack{i>j\\ i,j=1}}^{N-1}\lambda_i\lambda_j + \cdots + (-1)^{N-1}\prod_{i=1}^{N-1}\lambda_i\\ &= 1-\rho(\lambda)+\rho_2(\lambda)+\cdots+(-1)^{N-1}\rho_{N-1}(\lambda),\end{aligned}$$

where for $i = 1, \ldots, N-1$, ρ_i is the ith elementary symmetric polynomial in $\lambda_1, \ldots, \lambda_{N-1}$. We denote by $Q_i(\lambda_1, \ldots, \lambda_N)$ the ith elementary symmetric polynomials in $\lambda_1, \ldots, \lambda_N$, for $i = 1, \ldots, N-1$. By Ejov *et al.* [32, Theorem 2], for any Hamiltonian policy $\boldsymbol{f}_{\mathrm{H}}$, the characteristic polynomial $c(x)$ of $\mathbf{P}(\boldsymbol{f}_{\mathrm{H}})$ is $1 - x^N$. Hence, for $i = 1, \ldots, N$, all eigenvalues λ_i of $\mathbf{P}(\boldsymbol{f}_{\mathrm{H}})$ are the nth roots of unity. This implies that

$$Q_1(\lambda_1,\ldots,\lambda_N) = Q_2(\lambda_1,\ldots,\lambda_N) = \cdots = Q_{N-1}(\lambda_1,\ldots,\lambda_N) = 0.$$

Define $\rho_0(\lambda_1, \ldots, \lambda_{N-1}) = Q_0(\lambda_1, \ldots, \lambda_N) = 1$. As $\lambda_N = 1$, we have

$$\begin{aligned}\rho_1(\lambda_1,\ldots,\lambda_{N-1}) &= Q_1(\lambda_1,\ldots,\lambda_N) - \rho_0(\lambda_1,\ldots,\lambda_{N-1}),\\ \rho_2(\lambda_1,\ldots,\lambda_{N-1}) &= Q_2(\lambda_1,\ldots,\lambda_N) - \rho_1(\lambda_1,\ldots,\lambda_{N-1}),\\ &\vdots\\ \rho_{N-1}(\lambda_1,\ldots,\lambda_{N-1}) &= Q_{N-1}(\lambda_1,\ldots,\lambda_N) - \rho_{N-2}(\lambda_1,\ldots,\lambda_{N-1}),\end{aligned}$$

which gives us the result. □

Example 5.1 *We revisit the envelope graph (Figure 5.2) and consider two Hamiltonian cycles $\boldsymbol{f}_{\mathrm{H}_1}$ and $\boldsymbol{f}_{\mathrm{H}_2}$ on the graph, depicted respectively in Figures 5.3 and 5.4. The probability transition matrices induced by these policies are*

$$\mathbf{P}(\boldsymbol{f}_{\mathrm{H}_1}) = \begin{bmatrix} \cdot & 1 & \cdot & \cdot & \cdot & \cdot \\ \cdot & \cdot & \cdot & \cdot & \cdot & 1 \\ \cdot & \cdot & \cdot & 1 & \cdot & \cdot \\ \cdot & \cdot & \cdot & \cdot & 1 & \cdot \\ 1 & \cdot & \cdot & \cdot & \cdot & \cdot \\ \cdot & \cdot & 1 & \cdot & \cdot & \cdot \end{bmatrix} \quad \text{and} \quad \mathbf{P}(\boldsymbol{f}_{\mathrm{H}_2}) = \begin{bmatrix} \cdot & 1 & \cdot & \cdot & \cdot & \cdot \\ \cdot & \cdot & 1 & \cdot & \cdot & \cdot \\ \cdot & \cdot & \cdot & \cdot & \cdot & 1 \\ 1 & \cdot & \cdot & \cdot & \cdot & \cdot \\ \cdot & \cdot & \cdot & 1 & \cdot & \cdot \\ \cdot & \cdot & \cdot & \cdot & 1 & \cdot \end{bmatrix}.$$

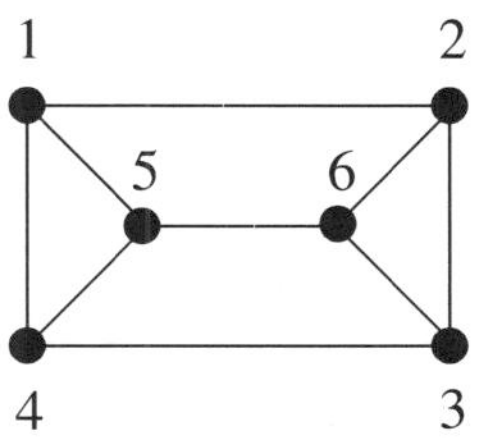

Fig. 5.2: A cubic graph G of order 6

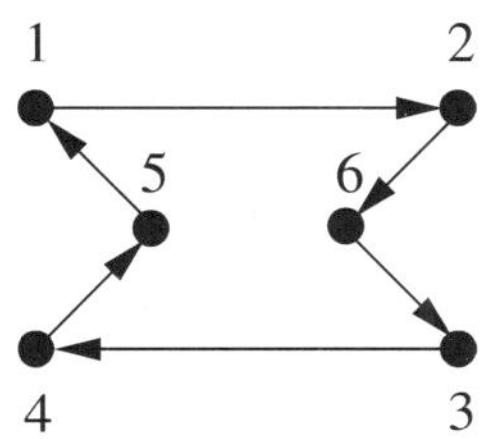

Fig. 5.3: The Hamiltonian cycle $\boldsymbol{f}_{H_1}$.

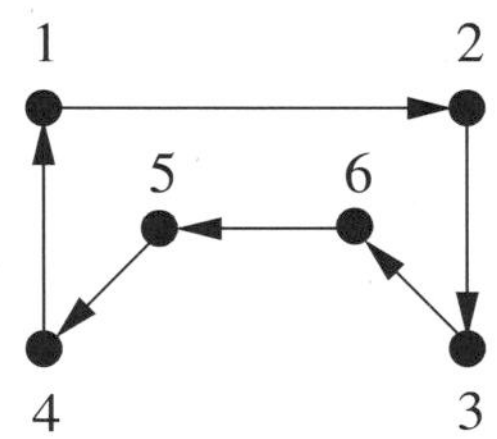

Fig. 5.4: The Hamiltonian cycle $\boldsymbol{f}_{H_2}$.

In this example, $N = 6$, *and* $\det \mathbf{W}(\boldsymbol{f}_{\mathrm{H}_1}) = \det \mathbf{W}(\boldsymbol{f}_{\mathrm{H}_2}) = 6$.

Proposition 5.2. *For any deterministic policy* $\boldsymbol{f} \in \mathcal{F}_{\mathcal{D}_1}$ *that contains a single cycle of length* $k < N$ *on a given graph of order* N,

$$\det \mathbf{W}(\boldsymbol{f}) = k. \tag{5.18}$$

Proof. Without loss of generality, we assume that $\mathbf{P}(\boldsymbol{f})$ contains the cycle $1 \to 2 \to \cdots \to k \to 1$, and that for $i \geq k+1$, $\mathbf{P}_{ij} = 1$ for some $j \in \{i+1\} \cup \{1, \ldots, k\}$, and $\mathbf{P}_{i\ell}(\boldsymbol{f}) = 0$ for all $\ell \neq j$. Hence, the $N \times N$ matrix $\mathbf{P}(\boldsymbol{f})$ is of the generic form

$$\left[\begin{array}{cccc|cccc} 0 & 1 & & & & & & \\ & \ddots & \ddots & & & \mathbf{0} & & \\ & & \ddots & 1 & & & & \\ 1 & & & 0 & & & & \\ \hline & & 1 & & 0 & 0 & & \\ & & & & & 0 & 1 & \\ & & & & & & \ddots & \\ & 1 & & & & & & 0 \end{array}\right]. \tag{5.19}$$

Let $\mathbf{E} = \{\boldsymbol{e}_1, \ldots, \boldsymbol{e}_N\}$ be the standard basis of $\mathbb{R}^N$ where for each $\boldsymbol{e}_i$, the ith entry is unity and all other elements are zeroes. Consider a new basis

$\mathbf{V} = \{\boldsymbol{v}_1, \ldots, \boldsymbol{v}_N\}$ of $\mathbb{R}^N$, with $\boldsymbol{v}_k = \boldsymbol{e}$ and

$$\boldsymbol{v}_i = \begin{cases} \boldsymbol{e}_{i+1} - \boldsymbol{e}_i & \text{for } i = 1, \ldots, k-1, \\ \boldsymbol{e}_i & \text{for } i = k+1, \ldots, N. \end{cases}$$

We define $\mathbf{B}(\boldsymbol{f}) = -\mathbf{P}(\boldsymbol{f}) + 1/N\mathbf{J}$, and simply write $\mathbf{B}$ where no ambiguity can arise. First, we show that vectors $\boldsymbol{v}_k, \boldsymbol{v}_{k+1}, \ldots, \boldsymbol{v}_N$ form a basis of the null space of $\mathbf{B}^{N-k+1}$, denoted by $\ker(\mathbf{B}^{N-k+1})$. It is clear that $\boldsymbol{v}_k = \boldsymbol{e} \in \ker(\mathbf{B})$, and $\mathbf{J}\boldsymbol{u} = \mu\boldsymbol{e}$, for any $\boldsymbol{u} \in \mathbb{R}^N$ and some $\mu \in \mathbb{R}$, since $\text{rank}(\mathbf{J}) = 1$ and $\mathbf{J}\boldsymbol{e} = \boldsymbol{e}$. Therefore, by (5.19),

$$\begin{aligned} \mathbf{B}\boldsymbol{v}_{k+1} &= -\mathbf{P}(\boldsymbol{f})\boldsymbol{e}_{k+1} + 1/N\mathbf{J}\boldsymbol{e}_{k+1} \\ &= 1/N\boldsymbol{e}, \end{aligned}$$

Similarly, for $k+2 \le i \le N$ and for $\gamma_i \in \{-1, 0\}$,

$$\mathbf{B}\boldsymbol{v}_i = \gamma_i \boldsymbol{v}_{i-1} + 1/N\boldsymbol{e},$$

If we consider $\mathbf{B}^m \boldsymbol{v}_i$, then (modulo multiple of $\boldsymbol{e}$) the longest possible chain could be

$$\boldsymbol{v}_N \overset{\mathbf{B}}{\to} \boldsymbol{v}_{N-1} \overset{\mathbf{B}}{\to} \boldsymbol{v}_{N-2} \overset{\mathbf{B}}{\to} \cdots \overset{\mathbf{B}}{\to} \boldsymbol{v}_{k+1} \overset{\mathbf{B}}{\to} \mathbf{0}.$$

At the last step, any multiple of $\boldsymbol{e}$ also vanishes. Thus, when restricted to the span of basic vectors $\{\boldsymbol{v}_k, \ldots, \boldsymbol{v}_N\}$, denoted by $\text{span}\{\boldsymbol{v}_k, \ldots, \boldsymbol{v}_N\}$, $\mathbf{B}$ is a nilpotent linear operator $\widetilde{\mathbf{B}}$, with $\widetilde{\mathbf{B}}^{N-k+1} = \mathbf{0}$. Hence, zero is an eigenvalue of $\mathbf{B}$ of multiplicity at least $N-k+1$. As $\boldsymbol{v}_i$ is an eigenvector of $\mathbf{I}$ corresponding to eigenvalue 1 for $i = k, \ldots, N$, we have

$$\det \mathbf{W}(\boldsymbol{f}) = \prod_{i=1}^{k-1} (1 - \lambda_i), \tag{5.20}$$

where λ_i's are the eigenvalues of the $(k-1) \times (k-1)$ leading principal minor of $\mathbf{B}$ in the $\mathbf{V}$ basis.

It remains to show that $\lambda_1, \ldots, \lambda_{k-1}$ coincide with eigenvalues of a $k \times k$ probability transition matrix $\mathbf{P}_{\text{H}_k}$ of a standard Hamiltonian cycle of length k with eigenvalue 1 excluded. Indeed, for $i = 1, \ldots, k-1$, $\mathbf{J}\boldsymbol{v}_i = 0$. Hence,

$$\mathbf{B}\boldsymbol{v}_1 = -\mathbf{P}(\boldsymbol{f})(\boldsymbol{e}_2 - \boldsymbol{e}_1) = -\boldsymbol{e}_1 + \boldsymbol{e}_k + \boldsymbol{v}_1^*,$$

where $\boldsymbol{v}_1^* \in \text{span}\ \{\boldsymbol{v}_k, \ldots, \boldsymbol{v}_N\}$. Analogously, for $i = 2, \ldots, k-1$,

$$\mathbf{B}\boldsymbol{v}_i = -\mathbf{P}(\boldsymbol{f})(\boldsymbol{e}_{i+1} - \boldsymbol{e}_1) = -\boldsymbol{e}_i + \boldsymbol{e}_k + \boldsymbol{v}_i^*,$$

where $v_i^* \in \text{span}\{\boldsymbol{v}_k, \ldots, \boldsymbol{v}_N\}$.

Therefore, we observe that the $(k-1) \times (k-1)$ leading principal minor for $\mathbf{B}$, namely $[\mathbf{B}]_{k-1}$, is the same as the principal leading minor $[\mathbf{M}]_{k-1}$ of $-\mathbf{P}_{\text{H}k} + 1/k\mathbf{J}_k$, if we choose bases $\mathbf{V}$ and $\{\boldsymbol{e}_l^{(k)} - \boldsymbol{e}_1^{(k)}, \ldots, \boldsymbol{e}_{k-1}^{(k)} - \boldsymbol{e}_1^{(k)}, \boldsymbol{e}_k\}$ respectively, where $\mathbf{J}_k$ is the $k \times k$ matrix of units, $\boldsymbol{e}_j^{(k)}$ is the jth vector of the standard unit basis of $\mathbb{R}^k$ and $\boldsymbol{e}^{(k)} = (1, 1, \ldots, 1)^{\text{T}} \in \mathbb{R}^k$.

Hence, for $r > 0$, principal minors of powers of $\mathbf{B}^r$ and $\mathbf{M}^r$ coincide and so the eigenvalues $\lambda_1, \ldots, \lambda_{k-1}$ of $[\mathbf{B}]_{k-1}$ and of $\mathbf{P}_{\text{H}_k}$, excluding eigenvalue 1, also coincide. By Proposition 5.1 with $N = k$, for $\mathbf{P}_{\text{H}_k}$,

$$\prod_{i=1}^{k-1}(1 - \lambda_i) = k, \tag{5.21}$$

which, together with (5.20), proves (5.18). □

Example 5.2 *We consider two deterministic policies $\boldsymbol{f}_1$ and $\boldsymbol{f}_2$ on the envelope graph, each policy possessing a cycle of length* 3 *and of length* 4*, respectively, depicted in Figures 5.5 and 5.6.*

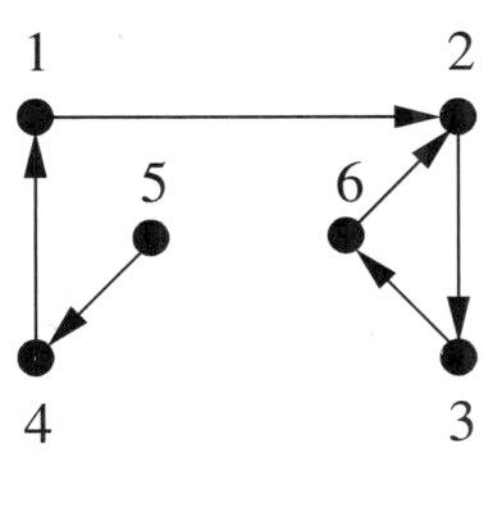

Fig. 5.5: Deterministic policy $\boldsymbol{f}_1$

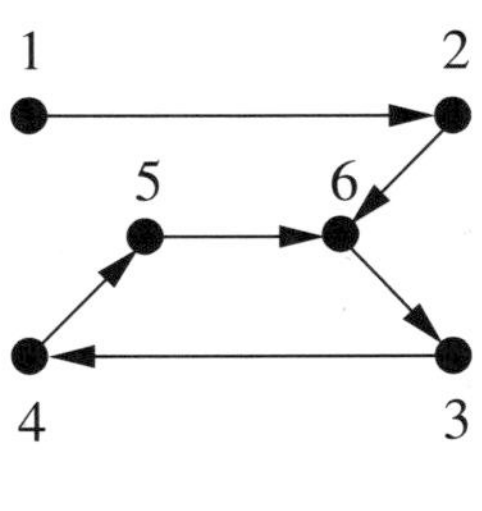

Fig. 5.6: Deterministic policy $\boldsymbol{f}_2$

The probability transition matrices induced by these two policies are

$$\mathbf{P}(\boldsymbol{f}_1) = \begin{bmatrix} \cdot & 1 & \cdot & \cdot & \cdot & \cdot \\ \cdot & \cdot & 1 & \cdot & \cdot & \cdot \\ \cdot & \cdot & \cdot & \cdot & \cdot & 1 \\ 1 & \cdot & \cdot & \cdot & \cdot & \cdot \\ \cdot & \cdot & \cdot & 1 & \cdot & \cdot \\ \cdot & 1 & \cdot & \cdot & \cdot & \cdot \end{bmatrix} \quad \textit{and} \quad \mathbf{P}(\boldsymbol{f}_2) = \begin{bmatrix} \cdot & 1 & \cdot & \cdot & \cdot & \cdot \\ \cdot & \cdot & \cdot & \cdot & \cdot & 1 \\ \cdot & \cdot & \cdot & 1 & \cdot & \cdot \\ \cdot & \cdot & \cdot & \cdot & 1 & \cdot \\ \cdot & \cdot & \cdot & \cdot & \cdot & 1 \\ \cdot & \cdot & 1 & \cdot & \cdot & \cdot \end{bmatrix}.$$

Simple calculations show that $\det \mathbf{W}(\boldsymbol{f}_1) = 3$, *and* $\det \mathbf{W}(\boldsymbol{f}_2) = 4$.

Proposition 5.3. *For any deterministic policy $\boldsymbol{f} \in \mathcal{F}_{\mathcal{D}_2}$ that contains $m \geq 2$ disjoint cycles of lengths k_i, $0 < k_i < N$ for $i = 1, \ldots, m$, on a given graph* *of order N,*

$$\det \mathbf{W}(\boldsymbol{f}) = 0. \tag{5.22}$$

Proof. As the policy $\boldsymbol{f}$ traces out two or more disjoint cycles, $\mathbf{P}(\boldsymbol{f})$ is reducible and consequently has an eigenvalue of unity with multiplicity two or more. By Corollary 5.2, equality (5.22) follows immediately. □

Example 5.3 *We consider two policies $\boldsymbol{f}_1$ and $\boldsymbol{f}_2$, each possessing two disjoint cycles: $\boldsymbol{f}_1$ has cycles $(1,4,5,1)$ and $(2,6,3,2)$, whereas $\boldsymbol{f}_2$ has cycles $(1,2,6,5,1)$ and $(3,4,3)$.*

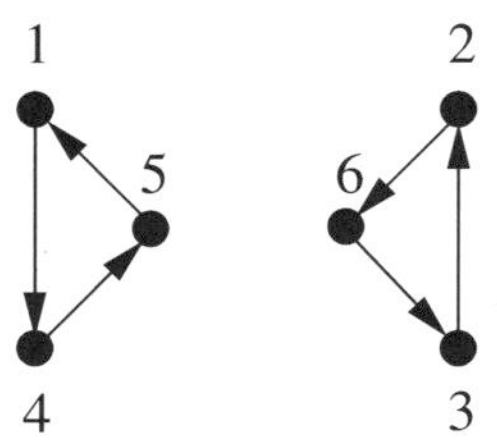

Fig. 5.7: Deterministic policy $\boldsymbol{f}_1 \in \mathcal{F}_{\mathcal{D}_2}$

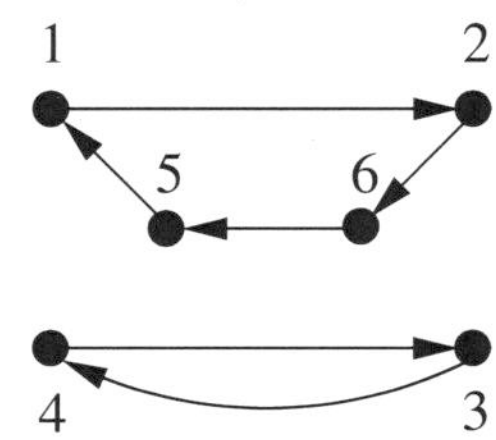

Fig. 5.8: Deterministic policy $\boldsymbol{f}_2 \in \mathcal{F}_{\mathcal{D}_2}$

The probability transition matrices induced by these two policies are

$$\mathbf{P}(\boldsymbol{f}_1) = \begin{bmatrix} \cdot & \cdot & \cdot & 1 & \cdot & \cdot \\ \cdot & \cdot & \cdot & \cdot & \cdot & 1 \\ \cdot & 1 & \cdot & \cdot & \cdot & \cdot \\ \cdot & \cdot & \cdot & \cdot & 1 & \cdot \\ 1 & \cdot & \cdot & \cdot & \cdot & \cdot \\ \cdot & \cdot & 1 & \cdot & \cdot & \cdot \end{bmatrix} \quad \text{and} \quad \mathbf{P}(\boldsymbol{f}_2) = \begin{bmatrix} \cdot & 1 & \cdot & \cdot & \cdot & \cdot \\ \cdot & \cdot & \cdot & \cdot & \cdot & 1 \\ \cdot & \cdot & \cdot & 1 & \cdot & \cdot \\ \cdot & \cdot & 1 & \cdot & \cdot & \cdot \\ 1 & \cdot & \cdot & \cdot & \cdot & \cdot \\ \cdot & \cdot & \cdot & \cdot & 1 & \cdot \end{bmatrix},$$

and, consequently, $\det \mathbf{W}(\boldsymbol{f}_1) = \det \mathbf{W}(\boldsymbol{f}_2) = 0$.

Proposition 5.4. *For any randomised policy $\boldsymbol{f} \in \mathcal{F}_\mathcal{S} \setminus \mathcal{F}_\mathcal{D}$ on a given graph of order N, there exist two deterministic policies $\bar{\boldsymbol{f}}, \hat{\boldsymbol{f}} \in \mathcal{F}_\mathcal{D}$ such that*

$$\det \mathbf{W}(\bar{\boldsymbol{f}}) \le \det \mathbf{W}(\boldsymbol{f}) \le \det \mathbf{W}(\hat{\boldsymbol{f}}). \tag{5.23}$$

Proof. We define the number of randomisations in rows of $\mathbf{P}(\boldsymbol{f})$ to be the number of elements $p_{ij} \neq \{0,1\}$. As $\boldsymbol{f} \notin \mathcal{F}_\mathcal{D}$, there exists at least one row—say, the first row—in $\mathbf{P}(\boldsymbol{f})$ such that a randomisation occurs in that row. This row has the structure

$$[\cdots\ a \cdots\ b \cdots\ c \cdots], \tag{5.24}$$

where $0 < a, b < 1$. While the argument is almost identical for the case of $a + b + c < 1$, for the sake of simplicity and without loss of generality, we assume $a+b+c=1$. Construct a feasible policy $\boldsymbol{d}_\nu$ such that $\mathbf{P}(\boldsymbol{d}_\nu)$ coincides with $\mathbf{P}(\boldsymbol{f})$ in rows $2,\dots,N$, and the first row of $\mathbf{P}(\boldsymbol{d}_\nu)$ is

$$[\cdots\ \nu\ \cdots\ \frac{(1-\nu)b}{1-a}\ \cdots\ \frac{(1-\nu)c}{1-a}\ \cdots], \tag{5.25}$$

where $0 \le \nu \le 1$. For $\nu = a$, $\mathbf{P}(\boldsymbol{d}_\nu) = \mathbf{P}(\boldsymbol{f})$ and $\boldsymbol{d}_\nu$ reduces to $\boldsymbol{f}$. Furthermore, $(1-\nu)b/(1-a), (1-\nu)c/(1-a) \in [0,1]$ because $b, c \le 1-a = b+c$. As $\det \mathbf{W}(\boldsymbol{d}_\nu)$ is a linear function in $\nu \in [0,1]$, we must have that either

- the maximum or the minimum of $\det \mathbf{W}(\boldsymbol{d}_\nu)$ occurs at $\boldsymbol{d}_0$ (or $\boldsymbol{d}_1$), and the other occurs at $\boldsymbol{d}_1$ (or $\boldsymbol{d}_0$), or
- $\det \mathbf{W}(\boldsymbol{d}_\nu) = \det \mathbf{W}(\boldsymbol{f})$ for all $\nu \in [0,1]$.

By (5.25), the first row of $\mathbf{P}(\boldsymbol{d}_0)$ reduces to

$$[\cdots\ 0\ \cdots\ \frac{b}{1-a}\ \cdots\ \frac{c}{1-a}\ \cdots], \tag{5.26}$$

and the first row of $\mathbf{P}(\boldsymbol{d}_1)$ reduces to

$$\left[\cdots\ 1\ \cdots\ 0\ \cdots\ 0\ \cdots\right]. \tag{5.27}$$

Define

$$\bar{\boldsymbol{d}} = \arg\min\{\det \mathbf{W}(\boldsymbol{d}_0), \det \mathbf{W}(\boldsymbol{d}_1)\}, \tag{5.28}$$

and

$$\hat{\boldsymbol{d}} = \arg\max\{\det \mathbf{W}(\boldsymbol{d}_0), \det \mathbf{W}(\boldsymbol{d}_1)\}. \tag{5.29}$$

As both matrices $\mathbf{P}(\bar{\boldsymbol{d}})$ and $\mathbf{P}(\hat{\boldsymbol{d}})$ coincide with $\mathbf{P}(\boldsymbol{f})$ in rows $2,\dots,N$ and their first row has the structure of either (5.26) or (5.27), both of $\mathbf{P}(\bar{\boldsymbol{d}})$ and $\mathbf{P}(\hat{\boldsymbol{d}})$ have at least one more zero in their rows than $\mathbf{P}(\boldsymbol{f})$. Furthermore, $\det \mathbf{W}(\bar{\boldsymbol{d}}) \le \det \mathbf{W}(\boldsymbol{f})$ and $\det \mathbf{W}(\hat{\boldsymbol{d}}) \ge \det \mathbf{W}(\boldsymbol{f})$. By induction on the number of randomisations, we obtain two deterministic policies $\bar{\boldsymbol{f}}, \hat{\boldsymbol{f}} \in \mathcal{F}_{\mathcal{D}}$ such that (5.23) holds. □

Proof of Theorem 5.2. Equations (5.17), (5.18) and (5.22) and Proposition 5.4 imply Theorem 5.1, which gives an upper bound for the matrix function $\det \mathbf{W}(\boldsymbol{f})$ for $\boldsymbol{f} \in \mathcal{F}_{\mathcal{S}}$. □

Corollary 5.3. *For any given graph G, the following statements hold:*

(i) If G is Hamiltonian, then

$$\max_{\boldsymbol{f} \in \mathcal{F}_{\mathcal{S}}} \det \mathbf{W}(\boldsymbol{f}) = N, \tag{5.30}$$

and $\det \mathbf{W}(\boldsymbol{f}) = N$ *if and only if* $\boldsymbol{f}$ *is a Hamiltonian policy.*

(ii) If G is non-Hamiltonian, then

$$\max_{\boldsymbol{f} \in \mathcal{F}_S} \det \mathbf{W}(\boldsymbol{f}) \leq N - 1. \tag{5.31}$$

Proof. (i) Equality (5.30) follows immediately from Theorem 5.2 as $\boldsymbol{f} \in \mathcal{F}_{\mathcal{H}}$ is a longest cycle on G. By Proposition 5.1, if $\boldsymbol{f} \in \mathcal{F}_{\mathcal{H}}$, then $\det \mathbf{W}(\boldsymbol{f}) = N$. Now it suffices to show that if $\det \mathbf{W}(\boldsymbol{f}) = N$, then $\boldsymbol{f}$ is a Hamiltonian policy.

Suppose $\det \mathbf{W}(\boldsymbol{f}) = N$ for some non-Hamiltonian policy $\boldsymbol{f}$. Define

$$\mathcal{T} = \{\boldsymbol{h} \in \mathcal{F}_S | \det \mathbf{W}(\boldsymbol{h}) = \max_{\boldsymbol{g} \in \mathcal{F}} \det \mathbf{W}(\boldsymbol{g}) = N\},$$

then $\boldsymbol{f} \in \mathcal{T}$. By Propositions 5.2 and 5.3, $\boldsymbol{f} \notin \mathcal{F}_D$. Therefore, there exists at least one row in $\mathbf{P}(\boldsymbol{f})$ where a randomisation occurs. Following an argument analogous to that presented in the proof of Proposition 5.4, there exists a reduction $\boldsymbol{f} \to \hat{\boldsymbol{d}}_1 \to \hat{\boldsymbol{d}}_2 \to \cdots \to \hat{\boldsymbol{d}}_r \to \hat{\boldsymbol{f}}$, such that $\hat{\boldsymbol{f}} \in \mathcal{F}_{\mathcal{D}}$ and

$$\det \mathbf{W}(\boldsymbol{f}) \leq \det \mathbf{W}(\hat{\boldsymbol{d}}_1) \leq \det \mathbf{W}(\hat{\boldsymbol{d}}_2) \leq \cdots \leq \det \mathbf{W}(\hat{\boldsymbol{d}}_r) \leq \det \mathbf{W}(\hat{\boldsymbol{f}}).$$

As

$$\det \mathbf{W}(\boldsymbol{f}) = \max_{\boldsymbol{g} \in \mathcal{F}} \det \mathbf{W}(\boldsymbol{g}) = N,$$

it follows that

$$\begin{aligned} \det \mathbf{W}(\boldsymbol{f}) &= \det \mathbf{W}(\hat{\boldsymbol{d}}_1) = \det \mathbf{W}(\hat{\boldsymbol{d}}_2) = \cdots = \det \mathbf{W}(\hat{\boldsymbol{d}}_r) \\ &= \det \mathbf{W}(\hat{\boldsymbol{f}}) = N. \end{aligned}$$

In the last stage of the reduction, that is, at $\hat{\boldsymbol{d}}_r \to \hat{\boldsymbol{f}}$, $\mathbf{P}(\hat{\boldsymbol{d}}_r)$ has exactly one row—say, the first row—where randomisation occurs, and the randomisation is on two edges. Then, the first row of $\mathbf{P}(\hat{\boldsymbol{d}}_r)$ is

$$[\cdots \ \nu \ \cdots \ \frac{(1-\nu)b}{1-a} \ \cdots],$$

where $a + b = 1$. As $\det \mathbf{W}(\hat{\boldsymbol{d}}_r) = N$ for all $\nu \in [0, 1]$, setting $\nu = 1$ and $\nu = 0$ gives us two distinct deterministic policies, one of which is $\hat{\boldsymbol{f}}$. Let the other be $\bar{\boldsymbol{f}}$, then both policies $\hat{\boldsymbol{f}}$ and $\bar{\boldsymbol{f}}$ must be distinct Hamiltonian cycles coinciding at exactly $N - 1$ edges, which is a contradiction.

(ii) This follows directly from Theorem 5.1, as the length of the longest cycle possible in a non-Hamiltonian graph is $N - 1$. □

Corollary 5.3 implies that there is a strictly positive gap between the maximum value of $\det \mathbf{W}(\boldsymbol{f})$ over the set of stochastic policies on a Hamiltonian graph and that on a non-Hamiltonian graph. Corollary 5.3 also gives an upper bound for a function of eigenvalues of any stochastic matrix. The following result follows immediately from Corollaries 5.2 and 5.3, by considering the set of feasible stochastic policies for a complete graph of order N that allows self-loops and setting $\varepsilon = 0$.

Corollary 5.4. *Let* $\mathbf{P}$ *be an* $N \times N$ *stochastic matrix, and denote by* $\lambda_1, \ldots, \lambda_{N-1}, \lambda_N = 1$ *the eigenvalues of* $\mathbf{P}$. *Then*

$$\prod_{i=1}^{N-1} (1 - \lambda_i) \leq N. \tag{5.32}$$

We note that complex eigenvalues of $\mathbf{P}$ can be rearranged in conjugate pairs, so the product in (5.32) is, indeed, a real number.

5.2.2 Perturbed Case

We restate Theorem 5.1 for $\varepsilon \in (0, 1)$ and $\alpha = 1/N$.

Theorem 5.3. *For any feasible policy* $\boldsymbol{f} \in \mathcal{F}_{\mathcal{S}}$ *on a given graph of order* N,

$$\det \mathbf{W}(\boldsymbol{f}, \varepsilon) \leq \frac{1 - (1 - \varepsilon)^k}{\varepsilon}, \tag{5.33}$$

where $k < N$ *is the circumference of the graph and* $\varepsilon \in (0, 1)$.

Similarly to that of Theorem 5.2, the proof of Theorem 5.3 requires the following results: Corollary 5.7 establishes a relationship between eigenvalues of $\mathbf{W}(\boldsymbol{f}, \varepsilon)$ and those of $\mathbf{P}^{\varepsilon}(\boldsymbol{f})$; Propositions 5.5–5.7 establish that (5.33) holds for a deterministic policy $\boldsymbol{f} \in \mathcal{F}_{\mathcal{H}}$, $\mathcal{F}_{\mathcal{D}_1}$ and $\mathcal{F}_{\mathcal{D}_2}$, respectively. Collectively, these results imply that (5.33) is true for any $\boldsymbol{f} \in \mathcal{F}_{\mathcal{D}}$; Proposition 5.8 shows that given a randomised policy $\boldsymbol{f} \in \mathcal{F}_{\mathcal{S}} \setminus \mathcal{F}_{\mathcal{D}}$, there exist two deterministic policies $\bar{\boldsymbol{f}}, \hat{\boldsymbol{f}} \in \mathcal{F}_{\mathcal{D}}$ such that

$$\det \mathbf{W}(\bar{\boldsymbol{f}}, \varepsilon) \leq \det \mathbf{W}(\boldsymbol{f}, \varepsilon) \leq \det \mathbf{W}(\hat{\boldsymbol{f}}, \varepsilon). \tag{5.34}$$

Inequalities (5.34), together with Propositions 5.5–5.7, indicate that (5.33) also holds for $\boldsymbol{f} \in \mathcal{F}_{\mathcal{S}} \setminus \mathcal{F}_{\mathcal{D}}$.

We denote by $\lambda_i^{\varepsilon}, \kappa_i^{\varepsilon}$ and μ_i^{ε} the eigenvalues of, respectively, the matrices $\mathbf{P}^{\varepsilon}(\boldsymbol{f})$, $\mathbf{P}^{\varepsilon}(\boldsymbol{f}) - 1/N\mathbf{J}$ and $\mathbf{W}(\boldsymbol{f}, \varepsilon)$, for $i = 1, \ldots, N$.

Remark 5.1. For $\varepsilon \in (0, 1)$, the matrix $\mathbf{P}^{\varepsilon}(\boldsymbol{f})$ is positive (elements $\mathbf{P}_{ij}^{\varepsilon}(\boldsymbol{f}) > 0$ for all i and j) and hence irreducible. Consequently, by the Perron-Frobenius theorem (see Horn and Johnson [66]), the eigenvalue 1 of $\mathbf{P}^{\varepsilon}(\boldsymbol{f})$ has multiplicity of 1.

Let $\mathbf{L} \in \mathbb{R}^{N\times N}$ be a stochastic matrix, $\mathbf{S} = \boldsymbol{e}\boldsymbol{v}^{\mathrm{T}} \in \mathbb{R}^{N\times N}$ be a rank-one stochastic matrix, where $\boldsymbol{v}$ is a probability distribution vector ($v_i > 0$ and $||\boldsymbol{v}||_1 = 1$) and $c \in (0,1)$. Serra-Capizzano [96] defines the matrix

$$\mathbf{L}(c) = c\mathbf{L} + (1-c)\mathbf{S}, \tag{5.35}$$

and proves the following result on the Jordan canonical form of $\mathbf{L}(c)$.

Theorem 5.4. [96] *Let ϖ_i be eigenvalues of $\mathbf{L}$ for $i = 1,\ldots,N$ and let $\mathbf{L} = \mathbf{X}\mathbf{T}(1)\mathbf{X}^{-1}$, where $\mathbf{X} = [\,\mathbf{e} \mid \boldsymbol{x}_2 \mid \cdots \mid \boldsymbol{x}_N\,], [\,\mathbf{X}^{-1}]^{\mathrm{T}} = [\,\mathbf{y}_1 \mid \mathbf{y}_2 \mid \cdots \mid \mathbf{y}_N\,]$,*

$$\mathbf{T}(c) = \begin{bmatrix} 1 & & & & \\ & c\varpi_2 & c* & & \\ & & \ddots & \ddots & \\ & & & c\varpi_{N-1} & c* \\ & & & & c\varpi_N \end{bmatrix} = \mathbf{D}^{-1} \begin{bmatrix} 1 & & & & \\ & c\varpi_2 & * & & \\ & & \ddots & \ddots & \\ & & & c\varpi_{N-1} & * \\ & & & & c\varpi_N \end{bmatrix} \mathbf{D},$$

with $\mathbf{D} = \mathrm{diag}(1, c, \ldots, c^{N-1})$, and $*$ denotes a value that can be 1 or 0 (and, consequently, $c*$ can either be c or 0). Then,

$$\mathbf{L}(c) = \mathbf{Z}\mathbf{T}(c)\mathbf{Z}^{-1}, \tag{5.36}$$

where $\mathbf{Z} = \mathbf{X}\mathbf{R}^{-1}$ and $\mathbf{R} = \mathbf{I} + \mathbf{e}_1\boldsymbol{w}^{\mathrm{T}}$, with $\boldsymbol{w} = (0, w_2, \ldots, w_N)^{\mathrm{T}}$, and

$$w_j = \begin{cases} \dfrac{(1-c)\boldsymbol{v}^{\mathrm{T}}\boldsymbol{x}_2}{1-c\varpi_2} & \text{for } j = 2, \\[2ex] \dfrac{[(1-c)\boldsymbol{v}^{\mathrm{T}}\boldsymbol{x}_j + [\mathbf{T}(c)]_{j-1,j}w_{j-1}]}{1-c\varpi_2} & \text{for } j = 3,\ldots,N. \end{cases}$$

Corollary 5.5. *For any feasible policy $\boldsymbol{f} \in \mathcal{F}_{\mathcal{S}}$ on a given graph of order N, the relationship between eigenvalues λ_i's of $\mathbf{P}(\boldsymbol{f})$ and eigenvalues λ_i^{ε}'s of $\mathbf{P}^{\varepsilon}(\boldsymbol{f})$ is*

$$\lambda_i^{\varepsilon} = \begin{cases} (1-\varepsilon)\lambda_i & \text{for } i = 1,\ldots,N-1, \\ 1 & \text{for } i = N. \end{cases} \tag{5.37}$$

Proof. As $\mathbf{P}^{\varepsilon}(\boldsymbol{f}) = (1-\varepsilon)\mathbf{P}(\boldsymbol{f}) + \varepsilon/N\mathbf{J}$, (5.37) follows from Theorem 5.4, by setting $c = 1-\varepsilon$, $\mathbf{L} = \mathbf{P}(\boldsymbol{f})$, $\mathbf{S} = 1/N\mathbf{J}$ and $\mathbf{L}(c) = \mathbf{P}^{\varepsilon}(\boldsymbol{f})$. Note that we assign the eigenvalue of 1 to λ_N and λ_N^{ε} (instead of λ_1 and λ_1^{ε}, as it would follow from the convention used in Theorem 5.4) for consistency throughout this chapter. □

Corollary 5.6. *For any feasible policy $\boldsymbol{f} \in \mathcal{F}_{\mathcal{S}}$ on a given graph of order N and for $i = 1,\ldots,N-1$, the eigenvalues λ_i^{ε}'s of $\mathbf{P}^{\varepsilon}(\boldsymbol{f})$ and eigenvalues κ_i^{ε}'s* *of* $\mathbf{P}^{\varepsilon}(\boldsymbol{f}) - 1/N\mathbf{J}$ *coincide, that is,*

$$\lambda_i^{\varepsilon} = \kappa_i^{\varepsilon}.$$

Furthermore, for $i = N$, $\lambda_N^{\varepsilon} = 1$ *corresponds to* $\kappa_N^{\varepsilon} = 0$.

Proof. First, it is simple to check that the proof for Theorem 5.4 still holds for $c > 1$. Using analogous arguments to those presented in the aforementioned proof, Lemma 5.6 follows. □

Corollary 5.7. *For any feasible policy* $\boldsymbol{f} \in \mathcal{F}_{\mathcal{S}}$ *on a given graph of order* N, *the eigenvalues* μ_i^{ε}*'s of* $\mathbf{W}(\boldsymbol{f})$ *are*

$$\mu_i^{\varepsilon} = \begin{cases} 1 - \lambda_i^{\varepsilon} & \text{for } i = 1, \ldots, N-1, \\ 1 & \text{for } i = N. \end{cases} \tag{5.38}$$

Proof. This result follows directly from Corollary 5.5 and Lemma 5.6. □

Corollary 5.8. *For any feasible policy* $\boldsymbol{f} \in \mathcal{F}_{\mathcal{S}}$ *on a given graph of order* N,

$$\det \mathbf{W}(\boldsymbol{f}, \varepsilon) = \prod_{i=1}^{N-1} (1 - \lambda_i^{\varepsilon}). \tag{5.39}$$

Proof. It is well-known that

$$\det \mathbf{W}(\boldsymbol{f}, \varepsilon) = \prod_{i=1}^{N} \mu_i^{\varepsilon}. \tag{5.40}$$

Then, Corollary 5.8 follows immediately from Corollary 5.7. □

Proposition 5.5. *For any Hamiltonian cycle* $\boldsymbol{f}_{\mathrm{H}} \in \mathcal{F}_{\mathcal{H}}$ *on a given graph of order* N *and for* $\varepsilon \in (0,1)$,

$$\det \mathbf{W}(\boldsymbol{f}_{\mathrm{H}}, \varepsilon) = \frac{1 - (1-\varepsilon)^N}{\varepsilon}. \tag{5.41}$$

Proof. By Corollaries 5.2 and 5.5, we obtain

$$\begin{aligned} &\det \mathbf{W}(\boldsymbol{f}_{\mathrm{H}}, \varepsilon) \\ &= \prod_{i=1}^{N-1} (1 - \lambda_i^{\varepsilon}) = \prod_{i=1}^{N-1} (1 - (1-\varepsilon)\lambda_i) \\ &= 1 - (1-\varepsilon) \sum_{i=1}^{N-1} \lambda_i + (1-\varepsilon)^2 \sum_{\substack{i>j \\ i,j=1}}^{N-1} \lambda_i \lambda_j - \cdots + (-1)^{N-1} (1-\varepsilon)^{N-1} \sum_{i=1}^{N-1} \lambda_i \end{aligned}$$

$$= 1 - (1-\varepsilon)q_1(\lambda_1,\dots,\lambda_{N-1}) + (1-\varepsilon)^2 q_2(\lambda_1,\dots,\lambda_{N-1})$$
$$- \dots + (-1)^{N-1}(1-\varepsilon)^{N-1} q_{N-1}(\lambda_1,\dots,\lambda_{N-1}), \tag{5.42}$$

where $q_i(\lambda_1,\dots,\lambda_{N-1})$ is the ith elementary symmetric polynomial in $\lambda_1,\dots,\lambda_{N-1}$ for $i = 1,\dots,N-1$. Let $Q_i(\lambda_1,\dots,\lambda_N)$ be the ith elementary symmetric polynomials in $\lambda_1,\dots,\lambda_N$, for $i = 1,\dots,N-1$. By Ejov *et al.* [32, Theorem 2], for any Hamiltonian policy $\boldsymbol{f}_{\mathrm{H}} \in \mathcal{F}_{\mathcal{H}}$, the characteristic polynomial $c(x)$ of $\mathbf{P}(\boldsymbol{f}_{\mathrm{H}})$ is $1 - x^N$. Hence, all eigenvalues λ_i of $\mathbf{P}(\boldsymbol{f}_{\mathrm{H}})$ are the nth roots of unity, for $i = 1,\dots,N$. This implies that

$$Q_1(\lambda_1,\dots,\lambda_N) = Q_2(\lambda_1,\dots,\lambda_N) = \dots = Q_{N-1}(\lambda_1,\dots,\lambda_N) = 0.$$

Define $q_0(\lambda_1,\dots,\lambda_{N-1}) = Q_0(\lambda_1,\dots,\lambda_N) = 1$. As $\lambda_N = 1$, we have

$$\begin{aligned}
q_1(\lambda_1,\dots,\lambda_{N-1}) &= Q_1(\lambda_1,\dots,\lambda_N) - q_0(\lambda_1,\dots,\lambda_{N-1}),\\
q_2(\lambda_1,\dots,\lambda_{N-1}) &= Q_2(\lambda_1,\dots,\lambda_N) - q_1(\lambda_1,\dots,\lambda_{N-1}),\\
&\vdots\\
q_{N-1}(\lambda_1,\dots,\lambda_{N-1}) &= Q_{N-1}(\lambda_1,\dots,\lambda_N) - q_{N-2}(\lambda_1,\dots,\lambda_{N-1}).
\end{aligned}$$

Consequently,

$$\begin{aligned}
q_1(\lambda_1,\dots,\lambda_{N-1}) &= -1,\\
q_2(\lambda_1,\dots,\lambda_{N-1}) &= 1,\\
&\vdots\\
q_{N-1}(\lambda_1,\dots,\lambda_{N-1}) &= (-1)^{N-1}.
\end{aligned}$$

Substituting these results into (5.42), we obtain

$$\det \mathbf{W}(\boldsymbol{f}_{\mathrm{H}},\varepsilon) = 1 + (1-\varepsilon) + (1-\varepsilon)^2 + \dots + (1-\varepsilon)^{N-1}.$$

□

Proposition 5.6. *For any deterministic policy $\boldsymbol{f} \in \mathcal{F}_{\mathcal{D}_1}$, that is, a policy possessing a single cycle of length $k < N$ on a given graph of order N and for $\varepsilon \in (0,1)$,*

$$\det \mathbf{W}(\boldsymbol{f},\varepsilon) = \frac{1-(1-\varepsilon)^k}{\varepsilon}. \tag{5.43}$$

Proof. As in the proof of Proposition 5.2, define $\mathbf{B}(\boldsymbol{f}) = -\mathbf{P}(\boldsymbol{f}) + 1/N\mathbf{J}$, and consider the $k \times k$ probability transition matrix $\mathbf{P}_{\mathrm{H}_k}$ of a standard Hamiltonian cycle of length k. For $i = 1,\dots,k-1$, let τ_i's be eigenvalues of $\mathbf{P}_{\mathrm{H}_k}$ excluding eigenvalue 1. By an argument presented in the proof of Proposition 5.1, τ_i's are the kth roots of unity. From the proof of Proposition 5.2, we have that eigenvalues ρ_i of $\mathbf{B}(\boldsymbol{f}) = -\mathbf{P}(\boldsymbol{f}) + 1/N\mathbf{J}$ are

$$\rho_i = \begin{cases} -\tau_i & \text{for } i = 1, \dots, k-1, \\ 0 & \text{for } i = k, \dots, N. \end{cases} \tag{5.44}$$

consequently, eigenvalues κ_i's of $\mathbf{P}(\boldsymbol{f}) - 1/N\mathbf{J}$ are

$$\kappa_i = \begin{cases} \tau_i & \text{for } i = 1, \dots, k-1, \\ 0 & \text{for } i = k, \dots, N. \end{cases} \tag{5.45}$$

By Corollary 5.6 and (5.45), eigenvalues λ_i's of $\mathbf{P}(\boldsymbol{f})$ are

$$\lambda_i = \begin{cases} \tau_i & \text{for } i = 1, \dots, k-1, \\ 0 & \text{for } i = k, \dots, N-1, \\ 1 & \text{for } i = N. \end{cases} \tag{5.46}$$

By Corollary 5.5 and (5.46), eigenvalues λ_i^ε's of $\mathbf{P}^\varepsilon(\boldsymbol{f})$ are

$$\lambda_i^\varepsilon = \begin{cases} (1-\varepsilon)\tau_i & \text{for } i = 1, \dots, k-1, \\ 0 & \text{for } i = k, \dots, N-1, \\ 1 & \text{for } i = N. \end{cases} \tag{5.47}$$

By Corollary 5.2 and (5.47),

$$\det \mathbf{W}(\boldsymbol{f}, \varepsilon) = \prod_{i=1}^{N-1} (1 - \lambda_i^\varepsilon) = \prod_{i=1}^{k-1} \{1 - (1-\varepsilon)\tau_i\}. \tag{5.48}$$

By the proof of Proposition 5.1,

$$\prod_{i=1}^{k-1} (1 - (1-\varepsilon)\tau_i) = \frac{1-(1-\varepsilon)^k}{\varepsilon}. \tag{5.49}$$

Consequently,

$$\det \mathbf{W}(\boldsymbol{f}, \varepsilon) = \frac{1-(1-\varepsilon)^k}{\varepsilon}.$$

□

Proposition 5.7. *For any deterministic policy* $\boldsymbol{f} \in \mathcal{F}_{\mathcal{D}_2}$, *that is, a policy possessing* $m \geq 2$ *disjoint cycles of lengths* k_i,

$$\det \mathbf{W}(\boldsymbol{f}, \varepsilon) = \varepsilon^{m-1} \prod_{i=1}^{m} \frac{1-(1-\varepsilon)^{k_i}}{\varepsilon}, \tag{5.50}$$

where $2 \leq k_i \leq N-2$ *and* $\sum_{i=1}^{m} k_i \leq N$.

Proof. For simplicity and without loss of generality, we assume that $m = 2$ and $\boldsymbol{f} \in \mathcal{F}_{\mathcal{D}_2}$ contains two disjoint cycles of lengths k_1, k_2 respectively, where $2 \le k_1, k_2 \le N - 2$ and $k_1 + k_2 \le N$. In particular, assume that

- the first cycle is $(1, 2, \ldots, k_1, 1)$, which we call the k_1-cycle, the second cycle is $((k_1 + 1), (k_1 + 2), \ldots, (k_1 + k_2 - 1), (k_1 + k_2), (k_1 + 1))$, which we call the k_2-cycle,
- for $i = k_1 + k_2 + 1$, $\boldsymbol{f}$ chooses to go to a vertex on either the k_1-cycle or the k_2-cycle, and for $i > k_1 + k_2 + 1 \in \mathcal{S}$, $\boldsymbol{f}$ either chooses to go to $i + 1$ or to a vertex on either the k_1-cycle or the k_2-cycle.

Hence, the $N \times N$ matrix $\mathbf{P}(\boldsymbol{f})$ is of the generic form

$$\mathbf{P}(\boldsymbol{f}) = \left[\begin{array}{c|c|c} \begin{matrix}0 & 1 & & \\ & \ddots & \ddots & \\ & & \ddots & 1 \\ 1 & & & 0\end{matrix} & \mathbf{0}_{k_1 \times k_2} & \mathbf{0}_{k_1 \times (N-(k_1+k_2))} \\ \hline \mathbf{0}_{k_2 \times k_1} & \begin{matrix}0 & 1 & & \\ & \ddots & \ddots & \\ & & \ddots & 1 \\ 1 & & & 0\end{matrix} & \mathbf{0}_{k_2 \times (N-(k_1+k_2))} \\ \hline \begin{matrix} & & \\ & & \\ & 1 & \end{matrix} & \begin{matrix} 1 & \\ & \\ & \end{matrix} & \begin{matrix}0 & 0 & & \\ & 0 & 1 & \\ & & \ddots & \\ & & & 0\end{matrix} \end{array}\right], \tag{5.51}$$

where $\mathbf{0}_{m \times n}$ is an $m \times n$ matrix of 0's.

For the ith cycle, $i = 1, 2$, we construct a vector $\boldsymbol{w}_i$ as follows

- every entry of $\boldsymbol{w}_i$ corresponding to a vertex on this cycle is 1,
- every entry corresponding to a vertex not on the cycle that is connected, by one or more edges, to this cycle is also 1, and
- every other entry is 0.

It is easy to verify that $\mathbf{P}\boldsymbol{w}_i = \boldsymbol{w}_i$, for $i = 1, 2$. Consequently, $\boldsymbol{w}_i$'s are eigenvectors of $\mathbf{P}$ corresponding to the eigenvalue 1. Moreover, it is clear that these eigenvectors are linearly independent and the multiplicity of the eigenvalue 1 of $\mathbf{P}$ is 2. Let n_i be the number of units in $\boldsymbol{w}_i$ for $i = 1, 2$. We define two linear combinations of $\boldsymbol{w}_1$ and $\boldsymbol{w}_2$ as follows

$$\boldsymbol{u}_1 = \boldsymbol{w}_2 - \frac{n_2}{n_1}\boldsymbol{w}_1, \quad \text{and} \quad \boldsymbol{u}_2 = \boldsymbol{w}_1 + \boldsymbol{w}_2 = \mathbf{e}.$$

With respect to the decomposition of $\mathbb{R}^N = \text{span}\{\mathbf{e}\} \bigoplus \ker(\mathbf{J})$, we define a new basis $\mathbf{V} = \{\boldsymbol{v}_1, \ldots, \boldsymbol{v}_N\}$ of $\mathbb{R}^N$ where

$$\boldsymbol{v}_i = \begin{cases} \mathbf{e}_{i+1} - \mathbf{e}_1 & \text{for } i = 1, \ldots, k_1 - 1, \\ \boldsymbol{u}_2 - \dfrac{n_2}{n_1}\boldsymbol{u}_1, & \text{for } i = k_1, \\ \mathbf{e}_{i+1} - \mathbf{e}_{k_1+1} & \text{for } i = k_1 + 1, \ldots, k_1 + k_2 - 1, \\ \mathbf{e} & \text{for } i = k_1 + k_2, \\ \mathbf{e}_i & \text{for } i = k_1 + k_2 + 1, \ldots, N. \end{cases} \tag{5.52}$$

We consider two standard Hamilonian cycles $\boldsymbol{f}_{\mathrm{H}k_1}$ and $\boldsymbol{f}_{\mathrm{H}k_1}$ of lengths k_1 and k_2, respectively, and their corresponding probability transition matrices $\mathbf{P}(\boldsymbol{f}_{\mathrm{H}k_1})$ and $\mathbf{P}(\boldsymbol{f}_{\mathrm{H}k_2})$. Following analogous arguments to those in the proof of Proposition 5.2, we can show that eigenvalues λ_i's of $\mathbf{P}(\boldsymbol{f})$ are

$$\lambda_i = \begin{cases} \bar{\tau}_i & \text{for } i = 1, \ldots, k_1 - 1, \\ 1 & \text{for } i = k, \\ \hat{\tau}_{i-k_1} & \text{for } i = k_1 + 1, \ldots, k_1 + k_2 - 1, \\ 0 & \text{for } i = k_1 + k_2, \ldots, N - 1, \\ 1 & \text{for } i = N, \end{cases} \tag{5.53}$$

where, for $i = 1, \ldots, k_1 - 1$ and for $j = 1, \ldots, k_2 - 1$, $\bar{\tau}_i$ and $\hat{\tau}_j$ are the eigenvalues of $\mathbf{P}(\boldsymbol{f}_{\mathrm{H}k_1})$ excluding eigenvalue 1 and $\mathbf{P}(\boldsymbol{f}_{\mathrm{H}k_2})$ excluding eigenvalue 1, respectively. By Corollary 5.5 and (5.53), eigenvalues λ_i^ε's of $\mathbf{P}^\varepsilon(\boldsymbol{f})$ are

$$\lambda_i^\varepsilon = \begin{cases} (1-\varepsilon)\bar{\tau}_i & \text{for } i = 1, \ldots, k_1 - 1, \\ 1 - \varepsilon & \text{for } i = k, \\ (1-\varepsilon)\hat{\tau}_{i-k_1} & \text{for } i = k_1 + 1, \ldots, k_1 + k_2 - 1, \\ 0 & \text{for } i = k_1 + k_2, \ldots, N - 1, \\ 1 & \text{for } i = N. \end{cases} \tag{5.54}$$

By Corollary 5.2 and (5.54), we obtain

$$\begin{aligned} \det \mathbf{W}(\boldsymbol{f}, \varepsilon) &= \prod_{i=1}^{N-1} (1 - \lambda_i^\varepsilon) \\ &= \varepsilon \prod_{i=1}^{k_1-1} (1 - \lambda_i^\varepsilon) \prod_{i=k_1+1}^{k_1+k_2-1} (1 - \lambda_i^\varepsilon) \\ &= \varepsilon \prod_{i=1}^{k_1-1} (1 - (1-\varepsilon)\bar{\tau}_i) \prod_{i=k_1+1}^{k_1+k_2-1} (1 - (1-\varepsilon)\hat{\tau}_{i-k_1}). \end{aligned}$$

By the proof of Proposition 5.1, we have

$$\prod_{i=1}^{k_1-1}(1-(1-\varepsilon)\bar{\tau}_i) = \frac{1-(1-\varepsilon)^{k_1}}{\varepsilon}, \tag{5.55}$$

and

$$\prod_{i=1}^{k_2-1}(1-(1-\varepsilon)\hat{\tau}_i) = \frac{1-(1-\varepsilon)^{k_2}}{\varepsilon}. \tag{5.56}$$

Consequently,

$$\det \mathbf{W}(\boldsymbol{f},\varepsilon) = \varepsilon\frac{1-(1-\varepsilon)^{k_1}}{\varepsilon}\frac{1-(1-\varepsilon)^{k_2}}{\varepsilon}. \tag{5.57}$$

Applying analogous arguments to those used above, we can generalise (5.57) to obtain (5.50) for $m > 2$. □

Proposition 5.8. *For any randomised policy $\boldsymbol{f} \in \mathcal{F}_{\mathcal{S}} \setminus \mathcal{F}_{\mathcal{D}}$, there exist two deterministic policies $\bar{\boldsymbol{f}}, \hat{\boldsymbol{f}} \in \mathcal{F}_{\mathcal{D}}$ such that*

$$\det \mathbf{W}(\bar{\boldsymbol{f}},\varepsilon) \le \det \mathbf{W}(\boldsymbol{f},\varepsilon) \le \det \mathbf{W}(\hat{\boldsymbol{f}},\varepsilon). \tag{5.58}$$

Proof. This proof follows completely analogous arguments to those presented in the proof of Proposition 5.4. □

Proof of Theorem 5.3. Assume $\varepsilon \in (0,1)$ and let L be the length of the longest cycle in the graph G. Obviously, $L = N$ if G is Hamiltonian. Equations (5.41), (5.43) and (5.50) and Proposition 5.8 imply that to prove (5.33), it suffices to show that

$$\frac{1-(1-\varepsilon)^L}{\varepsilon} > \frac{1-(1-\varepsilon)^k}{\varepsilon}, \tag{5.59}$$

and

$$\frac{1-(1-\varepsilon)^L}{\varepsilon} > \prod_{i=1}^{m}\varepsilon^{m-1}\frac{1-(1-\varepsilon)^{\ell_i}}{\varepsilon}, \tag{5.60}$$

where for $i = 1,\ldots,m$, $2 \le k, \ell_i \le L$, $m > 2$ and $\sum_{i=1}^{m} \ell_i \le N$.

As $k < L$, $(1-\varepsilon)^L < (1-\varepsilon)^k$ and we have (5.59). Furthermore, (5.60) also holds because

$$\prod_{i=1}^{m}\left(1-(1-\varepsilon)^{\ell_i}\right) \le 1-(1-\varepsilon)^{\ell_1} < 1-(1-\varepsilon)^L.$$

This completes the proof. □

In summary, in this chapter, we transformed HCP into an optimisation problem derived from results obtained in Chapters 3 and 4. Once again, we considered spaces of probability transition matrices of Markov chains induced by stationary policies of a Markov decision processes in which the underlying given graph G has been embedded as in Chapter 4. However, instead of fundamental matrices of these Markov chains we focussed on their negative generators corrected by an addition of a simple rank-one matrix (see (5.10)). We demonstrated that the determinant of such a matrix can act as a suitable functional for differentiating between Hamiltonian and non-Hamiltonian graphs by creating a similar, but stronger, Hamiltonicity gap as the variance of first hitting time functional discussed in Chapter 3 (compare Theorem 3.6 and Corollary 5.3). In the process, we proved that whenever the graph is Hamiltonian the maximal value of the above (unperturbed) determinant functional is simply equal to N, the order of the graph. Furthermore, we established desirable asymptotic properties of this functional under the symmetric linear perturbation first introduced in Section 3.2.

Chapter 6
Traces

6.1 Introduction

In this chapter, we present an alternative optimisation model that is equivalent to the Hamiltonian cycle problem. We show that instead of minimising top-left elements of the fundamental matrices like in Chapter 3, or maximising the determinant of the inverse of the fundamental matrices like in Chapter 5, one can minimise the traces of the fundamental matrices to determine the Hamiltonicity of a given graph. While it is unclear whether the trace function offers an algorithmic advantage over the top-left element or the determinant functions, the trace operator is certainly well studied and frequently employed. Moreover, this alternative objective function provides new probabilistic interpretations about the link between the Hamiltonian cycle problem and Markov chains.

6.2 Optimality at Hamiltonian Cycles

In this section, we consider a range of important properties of the trace of fundamental matrices—corresponding to Markov chains—induced by stationary policies in a Markov decision process into which the underlying graph has been embedded, as in Chapter 4. However, for simplicity of notation, we suppress the dependence of the probability transition matrices in question on the policies, as these do not play an explicit role in the derivations that follow. We discuss the properties of the trace functional separately for the cases when these transition matrices are unperturbed and perturbed according to the symmetric linear perturbation introduced in Section 3.2.

6.2.1 Unperturbed Case

We recall from Chapter 3 that the Cesaro-limit matrix $\mathbf{P}^*$ of a Markov chain with probability transition matrix $\mathbf{P}$ is given by

$$\mathbf{P}^*(\mathbf{P}) = \lim_{T\to\infty} \frac{1}{T+1} \sum_{t=0}^{T} \mathbf{P}^t,$$

and the fundamental matrix $\mathbf{G}(\mathbf{P})$ of the Markov chain is

$$\mathbf{G}(\mathbf{P}) = (\mathbf{I} - \mathbf{P} + \mathbf{P}^*)^{-1}.$$

Then, the deviation matrix $\mathbf{Z}(\mathbf{P})$ is defined as follows

$$\begin{aligned}\mathbf{Z}(\mathbf{P}) &= \mathbf{G}(\mathbf{P}) - \mathbf{P}\\ &= \lim_{T\to\infty} \frac{1}{T+1} \sum_{t_0=0}^{T} \sum_{t=0}^{t_0} (\mathbf{P}^t - \mathbf{P}^*). \end{aligned} \tag{6.1}$$

If $\mathbf{P}$ is aperiodic, we have

$$\mathbf{Z}(\mathbf{P}) = \sum_{t=0}^{\infty} (\mathbf{P}^t - \mathbf{P}^*). \tag{6.2}$$

When no ambiguity can arise, we simply write $\mathbf{Z}(\mathbf{P})$ as $\mathbf{Z}$. In Chapter 3, we denote by $\boldsymbol{q}$ the stationary distribution vector, by τ_1 the first hitting time of vertex 1, by $\mathbb{E}_i[\tau_1]$ the expectation of the first hitting time to vertex 1 after starting at $i \neq 1$, and by $\mathbb{E}_1[\tau_1]$ the expectation of the first return time to vertex 1 after starting from there. Generalising the notation, we denote by τ_i the first hitting time of vertex i, for $i \in \mathcal{S}$, by $\mathbb{E}_{\boldsymbol{q}}[\tau_i]$ the expectation of the first hitting time to vertex i after starting at a vertex determined by the stationary distribution $\boldsymbol{q}$, and by ρ_i the first hitting time to vertex i after starting at vertex i.

In the irreducible case, there is a close relationship between the diagonal elements of $\mathbf{Z}$ and hitting times τ_i. By Aldous and Fill [2, Lemma 11],

$$q_i \mathbb{E}_{\boldsymbol{q}}[\tau_i] = \mathbf{Z}_{ii}. \tag{6.3}$$

Combining it with formula (21) in the same chapter, we obtain

$$\begin{aligned}\mathbb{E}_i[\tau_i^2] &= 2\mathbb{E}_{\boldsymbol{q}}[\tau_i]/q_i + 1/q_i \\ &= 2\mathbf{Z}_{ii}/q_i^2 + 1/q_i.\end{aligned} \tag{6.4}$$

Let $c_i^2 = \mathbb{E}[\rho_i^2]/(\mathbb{E}[\rho_i])^2$ be the coefficient of variation of ρ_i. Since $\mathbb{E}[\rho_i] = 1/q_i$, it follows from (6.4) that

$$\begin{aligned}c_i^2 &= q_i^2 \mathbb{E}_i[\tau_i^2] \\ &= 2\mathbf{Z}_{ii} + q_i, \quad i = 1, \ldots, N.\end{aligned} \tag{6.5}$$

Summing over i, we get

$$\sum_{i=1}^{N} c_i^2 = 2\sum_{i=1}^{N} \mathbf{Z}_{ii} + 1, \tag{6.6}$$

and thus, minimising either of the two sums in (6.6) is equivalent. This gives us an alternative probabilistic proof of the next proposition, which has been recently obtained in Hunter [67] by linear algebraic methods.

Proposition 6.1. *On a set of irreducible transition matrices* $\mathbf{P}$ *and for any Hamiltonian cycle* $\mathbf{P}_{\mathrm{H}}$ *on a given graph,*

$$\min_{\mathbf{P}} \mathrm{Tr}[\mathbf{Z}(\mathbf{P})] = \mathrm{Tr}[\mathbf{Z}(\mathbf{P}_{\mathrm{H}})] = (N-1)/2.$$

Equivalently, $\sum_{i=1}^{N} c_i^2$ *achieves its minimal value* N *at matrix* $\mathbf{P}$ *if and only if* $\mathbf{P}$ *induces a Hamiltonian cycle.*

Proof. If $\mathbf{P}$ is a Hamiltonian matrix, then for any $i \in \mathcal{S}$, we have $\rho_i = N$ implying that c_i^2 takes its minimal possible value 1. Thus, the Hamiltonian cycle provides $\sum_i c_i^2 = N$ which is the minimal possible value of this sum. Moreover, no other irreducible matrix has such a property. Indeed, in any irreducible non-Hamiltonian Markov chain, there exists a state j such that ρ_j is not deterministic, implying $c_j^2 > 1$, and, consequently, $\sum_i c_i^2 > N$. □

We now generalise Proposition 6.1 to Markov chains with arbitrary structure. Assume that $\mathbf{P}$ is such that there is a class of transient states $\mathcal{C}_0$ and a class of ergodic states $\mathcal{C}_1$.

Proposition 6.2. *If* $\mathcal{C}_0$ *is a non-empty class of transient states and* $\mathcal{C}_1$ *is an irreducible class of ergodic states induced by the matrix* $\boldsymbol{P}$*, then*

$$\mathrm{Tr}[\mathbf{Z}] \geq |\mathcal{C}_0| + \frac{|\mathcal{C}_1| - 1}{2} > \frac{N-1}{2}.$$

Proof. For $i \in \mathcal{C}_0$, we have $q_i = 0$ and thus it follows from the definition of $\mathbf{Z}$ that $Z_{ii} \geq 1$. Furthermore, the transition probabilities restricted to $\mathcal{C}_1$ constitute a transition matrix of a Markov chain on $\mathcal{C}_1$. Note that the latter Markov chain is ergodic. Moreover, according to the definition, the deviation matrix

$\mathbf{Z}^{(1)}$ of this smaller Markov chain is the same as the corresponding block of the matrix $\mathbf{Z}$ of the original chain. Thus, it follows from Proposition 6.1 that

$$\sum_{i \in \mathcal{C}_1} \mathbf{Z}_{ii} = \mathrm{Tr}[\mathbf{Z}^{(1)}] \geq \frac{|\mathcal{C}_1| - 1}{2},$$

and the equality is achieved on a Hamiltonian cycle in the sub-graph whose vertices are from $\mathcal{C}_1$. Putting everything together, we get

$$\mathrm{Tr}[\mathbf{Z}] = \sum_{i \in \mathcal{C}_0} \mathbf{Z}_{ii} + \sum_{i \in \mathcal{C}_1} \mathbf{Z}_{ii} \geq |\mathcal{C}_0| + \frac{|\mathcal{C}_1| - 1}{2} > \frac{N-1}{2}$$

where the last inequality follows simply because $N = |\mathcal{C}_0| + |\mathcal{C}_1|$. □

It follows from the above proposition that the chains with one ergodic and one transient class of states can not be candidates for minimizing $\mathrm{Tr}[\mathbf{Z}]$. Now, consider a multi-chain with several ergodic classes. Without loss of generality, we unite all transient states in a single transient class. The following theorem is a generalization of [67, Theorem 4.2] and Proposition 6.1 to the class of all stochastic matrices.

Theorem 6.1. *On a set of all stochastic matrices, the objective function* $\mathrm{Tr}[\mathbf{G}]$ *achieves its minimal value* $(N+1)/2$ *at* $\mathbf{P}$ *if and only if* $\mathbf{P}$ *induces a Hamiltonian cycle.*

Proof. The argument goes as in the proof of Proposition 6.2. Assume that $\mathbf{P}$ induces a Markov chain with M ergodic classes of communicating states $\mathcal{C}_1, \ldots, \mathcal{C}_M$. For $m = 1, \ldots, M$, the transition probabilities restricted to $\mathcal{C}_m$ constitute a transition matrix of a Markov chain on $\mathcal{C}_m$. The latter smaller Markov chain is ergodic, and its deviation matrix $\mathbf{Z}^{(m)}$ is the same as the corresponding block of the matrix $\mathbf{Z}$ of the original chain. Moreover, the corresponding block $\mathbf{P}^{*(m)}$ of $\mathbf{P}^*$ is an ergodic projection of the Markov chain on $\mathcal{C}_m$. Thus, it follows from (6.5) that

$$\begin{aligned}
\mathrm{Tr}[\mathbf{Z}^{(m)} + \mathbf{P}^{*(m)}] &= \sum_{i \in \mathcal{C}_m} (\mathbf{Z}_{ii} + \mathbf{P}^{*(m)}_{ii}) \\
&= \frac{1}{2} \sum_{i \in \mathcal{C}_m} (c_i^2 + \mathbf{P}^{*(m)}_{ii}) \\
&= \frac{1}{2} \sum_{i \in \mathcal{C}_m} c_i^2 + \frac{1}{2}, \quad \text{for } m = 1, \ldots, M.
\end{aligned}$$

Now, summing over all ergodic classes and transient states and using Proposition 6.2 we obtain

$$\mathrm{Tr}[\mathbf{G}] = \mathrm{Tr}[\mathbf{Z} + \mathbf{P}^*] = \sum_{i \in \mathcal{C}_0} \mathbf{Z}_{ii} + \sum_{m=1}^{M} \sum_{i \in \mathcal{C}_m} [\mathbf{Z}_{ii} + \mathbf{P}^{*(m)}_{ii}]$$

$$= \sum_{i \in \mathcal{C}_0} \mathbf{Z}_{ii} + \frac{1}{2} \sum_{i=1}^{N} c_i^2 + \frac{1}{2} M \geq |\mathcal{C}_0| + \frac{N}{2} + \frac{M}{2} \geq \frac{N+1}{2}.$$

We see that if all coefficients of variation equal 1 then $\mathrm{Tr}[\mathbf{G}]$ is minimised if there is only one ergodic class and no transient state. Then, we are back to the irreducible case, and according to Proposition 6.1, the minimal value

$$(N-1)/2 + \sum_{i=1}^{N} \mathbf{P}^*_{ii} = (N-1)/2 + \sum_{i=1}^{N} q_i = (N+1)/2$$

is achieved if and only if $\mathbf{P}$ induces a Hamiltonian cycle. □

In Chapter 3, we formulated the Hamiltonian cycle problem as an optimisation problem, where the objective function is to minimise the first diagonal element $\mathbf{G}_{11}$ of a fundamental matrix, rather than the trace $\mathrm{Tr}[\mathbf{G}]$. The next proposition shows that these two criteria are equivalent on the set $\mathcal{F}_{\mathcal{DS}}$ of doubly stochastic Markov chains. We recall that for any doubly stochastic Markov chain, the stationary distribution vector $\boldsymbol{q}$ is uniform and thus there cannot be any transient state.

Proposition 6.3. *Both* $\mathbf{G}_{11}$ *and* $\mathrm{Tr}[\mathbf{G}]$ *attain their minima over doubly stochastic matrices at a Hamiltonian cycle when one exists.*

Proof. Since all states are recurrent, any state $i \in \mathcal{S}$ belongs to one of the disjoint ergodic classes $\mathcal{C}_1, \ldots, \mathcal{C}_M$, each of which can be seen as a separate doubly stochastic Markov chain. Then, assuming $i \in \mathcal{C}_1$, from (6.5) we obtain

$$\mathbf{G}_{ii} = \mathbf{Z}_{ii} + \mathbf{P}^*_{ii} = 1/2(c_i^2 + 1/|\mathcal{C}_1|),$$

which is clearly minimised on a Hamiltonian cycle with $c_i^2 = 1$ and $1/|\mathcal{C}_1| = 1/N$. Now, let $\mathbf{P}$ be doubly stochastic and irreducible so that $q_i = 1/N$, and assume that $\mathbf{P}$ minimises c_i^2, that is, $c_i^2 = 1$. Then, with probability 1, ρ_i equals its expectation N. Since ρ_i is deterministic, each state has to be visited exactly once on the way from i back to i. Moreover, the order of visiting the states must be fixed, otherwise, there will be a possibility to shorten or extend the cycle. Thus, we conclude that the condition $c_i^2 = 1$ in a doubly stochastic chain ensures the Hamiltonian cycle and, consequently, $c_j^2 = 1$ for all $j \in \mathcal{S}$. Hence, according to Proposition 6.1, $\mathrm{Tr}[\mathbf{G}] = \mathrm{Tr}[\mathbf{Z}] + 1$ is also minimised on this matrix. □

Consider an arbitrary irreducible stochastic matrix $\mathbf{P}$. Since $c_1^2 \geq 1$, it follows that

$$\mathbf{G}_{11} = \mathbf{Z}_{11} + q_1 = (c_1^2 + q_1)/2 > 1/2 + 1/(2N)$$

if $q_1 > 1/N$. Assume now that $q_1 < 1/N$. Then, the average cycle length between two successive visits to state 1 is greater than N, and thus the cycle typically has loops implying that the cycle length is rather variable. The question is whether the gain in q_1 can compensate for the increased coefficient of variation when minimising $\mathbf{G}_{11}$. We partially resolve this problem by considering a particular class of Markov chains that is described in Hunter [67] as *cycle drifts.* A *cycle drift* is defined as follows: at each state $i = 1, \ldots, N$, with probability p_i, the process follows a cyclic path $1 \to 2 \to \cdots \to N \to 1$. With complementary probability $1 - p_i$, the process makes a self-loop and stays in state i. A Hamiltonian cycle is a special case of a cycle drift, where a probability of a self-loop is zero in each state. The following proposition holds.

Proposition 6.4. *On a class of Markov chains inducing cycle drifts, the minimum of* $\mathbf{G}_{11}$ *is equal to* $1/2 + 1/(2N)$*, and is achieved on a Hamiltonian cycle.*

Proof. Assume that a cycle drift has a stationary distribution $\boldsymbol{q}$ with $q_1 \leq 1/N$. Define a renewal cycle as a time between two subsequent transitions from N to 1. We denote by C the average length of the renewal cycle. Since the length of stay in each state is geometric, we have

$$C = \sum_{i=1}^{N} 1/p_i \geq N.$$

We observe that the beginning of each renewal cycle constitutes a regenerative epoch of our Markov process; hence, the Renewal Theory [25] implies that for $i \in \mathcal{S}$,

$$q_i = \frac{\mathbb{E}[\text{time in state } i \text{ on a cycle}]}{\mathbb{E}[C]} = 1/(p_i C) \geq 1/C.$$

Now, we use the formula $\mathbf{Z}_{11} = q_1 \mathbb{E}_{\boldsymbol{q}}[\tau_1]$ to compute $\mathbf{Z}_{11}$. For $\mathbb{E}_{\boldsymbol{q}}[\tau_1]$, we obtain

$$\begin{aligned}
\mathbb{E}_{\boldsymbol{q}}[\tau_1] &= 0 \cdot q_1 + \sum_{i=2}^{N} q_i \sum_{j=i}^{N} \frac{1}{p_j} \\
&= \frac{1}{C} \sum_{i=2}^{N} \frac{1}{p_i} \sum_{j=i}^{N} \frac{1}{p_j} \\
&= \frac{1}{2C} (\sum_{i=2}^{N} 1/p_i)^2 + \frac{1}{2C} \sum_{i=2}^{N} (1/p_i)^2 \\
&= \frac{1}{2C} (C - 1/p_1)^2 + 1/(2C) \sum_{i=2}^{N} (1/p_i)^2. \qquad (6.7)
\end{aligned}$$

It is straightforward that the second term in the last equality is minimised when $1/p_i$ is constant for $i = 2, \ldots, N$. Then,

$$\frac{1}{p_i} = \frac{C - 1/p_1}{N - 1},$$

and (6.7) reduces to

$$\begin{aligned}\mathbb{E}_{\boldsymbol{q}}[\tau_1] &= \frac{1}{2C}\left(C - 1/p_1\right)^2 + \frac{1}{2C}\,\frac{(C-1/p_1)^2}{N-1} \\ &= \frac{C}{2}(1-q_1)^2\,\frac{N}{N-1}.\end{aligned}$$

Next, we determine which value of $C \geq N$ and which value of $q_1 \in [1/C, 1/N]$ minimise the function

$$\begin{aligned}\mathbf{Z}_{11} + q_1 &= q_1\mathbb{E}_{\pi}[\tau_1] + q_1 \\ &= \frac{C}{2}(1-q_1)^2 q_1\,\frac{N}{N-1} + q_1. \end{aligned} \tag{6.8}$$

For $N \geq 3$, it is easy to check that the expression on the right-hand side of (6.8) is increasing in q_1 for $q_1 \leq 1/N$ and thus, for minimising $\mathbf{Z}_{11} + q_1$ with given C, we have to choose the smallest value $q_1 = 1/C$. Substituting this into (6.8), we get

$$\mathbf{Z}_{11} + q_1 = \frac{1}{2}(1 - 1/C)^2\,\frac{N}{N-1} + 1/C,$$

which is minimised at $C = N$. For $N = 2$, the right-hand side of (6.8) achieves its minimal value when $q_1 = 1/2$ and $C = 2$. Thus, the $\mathbf{Z}_{11} + q_1$ is minimised when $p_i = 1$ for $i \in \mathcal{S}$, which corresponds to the Hamiltonian cycle. □

6.2.2 Perturbed Case

In Section 6.2.1, we showed that the Hamiltonian cycle problem can be solved by minimising $\mathrm{Tr}[\mathbf{G}] = \mathrm{Tr}[\mathbf{Z} + \mathbf{P}^*]$ on a class of all stochastic matrices corresponding to a given graph. While theoretically interesting, this formulation is not very practical for several reasons. First, the objective function is difficult to compute. There is no guarantee that series in (6.1) converge fast enough. Evaluating the matrix inverse $\mathbf{G}$ might be computationally demanding for large graphs, and it requires the knowledge of $\mathbf{P}^*$. Second, a deviation matrix turns out to be very sensitive to small changes in the transition matrix $\mathbf{P}$. Such sensitivity finds its explanation in the theory of singularly perturbed Markov chains. Recall that a perturbation of a Markov chain is *singular* if it alters the ergodic structure of the chain, for instance, if a reducible chain becomes irreducible, equivalently, some transient states become recurrent, or the number of ergodic classes changes. Assume for example that the transition matrix $\mathbf{P}$ has M ergodic classes. Then, by adding small but positive transition probabilities between two states in different classes, we change the number of classes which leads to drastic changes in the deviation matrix [7]. In the special case of symmetric linear perturbation, we observe this phenomenon in Proposition 6.5. For a comprehensive survey on singularly perturbed Markov chains and Markov decision processes, we refer to Avrachenkov *et al.* [5].

Given the sensitivity of the deviation matrix to singular perturbations, it would be convenient to preserve the ergodic structure of the chain while changing the transition probabilities. Ideally, we would like to keep the transition matrix irreducible but imposing irreducibility on $\mathbf{P}$ is inconvenient. A powerful way to guarantee irreducibility is to introduce a symmetric linear perturbation as in Chapters 3 and 5

$$\mathbf{P}^{\varepsilon} = (1-\varepsilon)\mathbf{P} + \varepsilon/N\mathbf{J}, \tag{6.9}$$

for $\varepsilon \in (0,1)$. The transition matrix $\mathbf{P}^{\varepsilon}$ induces a Markov chain that at each step follows $\mathbf{P}$ with probability $(1-\varepsilon)$ and picks the next state uniformly at random with probability ε. The symmetric perturbation $\varepsilon/N\mathbf{J}$ ensures irreducibility and aperiodicity of $\mathbf{P}^{\varepsilon}$. This implies that the corresponding deviation matrix $\mathbf{Z}(\varepsilon)$ is much less sensitive to changes in $\mathbf{P}$ than the original deviation matrix $\mathbf{Z}$. We note that since $\mathbf{P}^{\varepsilon}$ is irreducible, we have

$$\begin{aligned}\mathrm{Tr}[\mathbf{G}(\varepsilon)] &= \mathrm{Tr}[\mathbf{Z}(\varepsilon) + \mathbf{P}^*(\varepsilon)] \\ &= \mathrm{Tr}[\mathbf{Z}(\varepsilon)] + 1,\end{aligned} \tag{6.10}$$

where $\mathbf{G}(\varepsilon)$ is the fundamental matrix of $\mathbf{P}^{\varepsilon}$. In what follows, we operate with $\mathbf{Z}(\varepsilon)$ rather than $\mathbf{G}(\varepsilon)$ because $\mathrm{Tr}[\mathbf{Z}(\varepsilon)]$ has a natural probabilistic interpretation via the Random Target lemma [2, Chapter 2, Corollary 13]. The results obtained for $\mathrm{Tr}[\mathbf{Z}(\varepsilon)]$ apply to $\mathrm{Tr}[\mathbf{G}(\varepsilon)]$ straightforwardly via (6.10).

We observe that if $\mathbf{P}$ is reducible then the symmetric linear perturbation is singular. Thus, the results from perturbation analysis suggest that for any $i \in \mathcal{S}$, the quantity $\mathbf{Z}_{ii}(\varepsilon)$ tends to infinity as ε approaches 0. Therefore, when ε is small enough, all reducible chains have a higher value of $\mathrm{Tr}[\mathbf{Z}(\varepsilon)]$ than the perturbed Hamiltonian path. Hence, a small perturbation helps to filter out reducible matrices when minimising $\mathrm{Tr}[\mathbf{Z}(\varepsilon)]$. This argument is formalised in the next proposition.

Proposition 6.5.

(*i*) *If* $\mathbf{P}$ *is reducible, then*

$$\mathrm{Tr}[\mathbf{Z}(\varepsilon)] \geq \frac{1}{N\varepsilon}.$$

(*ii*) *If* $\mathbf{P}$ *induces a Hamiltonian cycle, then for all* $\varepsilon < 1/N$,

$$\mathrm{Tr}[\mathbf{Z}(\varepsilon)] \leq \frac{N-1}{2} + \frac{(N-1)^2}{2}\varepsilon. \tag{6.11}$$

Proof. To prove Part (i), note that a reducible matrix $\mathbf{P}$ induces at least one class $\mathcal{C}(\varepsilon)$ (transient or ergodic) such that in the perturbed chain,

$$\sum_{j \in \mathcal{C}(\varepsilon)} q_j(\varepsilon) \geq 1/N. \tag{6.12}$$

Now, choose some state $i \notin \mathcal{C}(\varepsilon)$ as an initial state. In order to reach $\mathcal{C}(\varepsilon)$ starting from i, at least one ε-transition has to be made, so the mean hitting time is not smaller than $1/\varepsilon$. Hence, according to the Random Target lemma [2, Chapter 2, Corollary 13],

$$\begin{aligned}\mathrm{Tr}[\mathbf{Z}(\varepsilon)] &= \sum_j q_j(\varepsilon)\mathbb{E}_i[\tau_j(\varepsilon)] \geq \sum_{j \in \mathcal{C}(\varepsilon)} q_j(\varepsilon)\mathbb{E}_i[\tau_j(\varepsilon)] \\ &\geq \frac{1}{\varepsilon} q_{\mathcal{C}(\varepsilon)}(\varepsilon) \geq \frac{1}{N\varepsilon}.\end{aligned}$$

For Part (ii), consider the perturbation of the matrix inducing a Hamiltonian cycle. Applying the Random Target lemma and taking into account that $q_j(\varepsilon) = 1/N$ because the perturbed matrix is doubly stochastic, we get

$$\mathrm{Tr}[\mathbf{Z}(\varepsilon)] = \sum_j q_j(\varepsilon)\mathbb{E}_1[\tau_j(\varepsilon)] = \frac{1}{N}\sum_j \mathbb{E}_1[\tau_j(\varepsilon)]. \tag{6.13}$$

Given that no ε-transition occurs on the way from 1 to j, the last expression of (6.13) equals $(N-1)/2$. The probability of at least one ε-transition on the way from 1 to j is at most $(1-(1-\varepsilon)^{N-1})$. Furthermore, after an ε-transition, the state j has to be reached starting from some new initial distribution $\boldsymbol{\mu}$.

Using again the Random Target lemma, we obtain

$$\begin{aligned}\mathrm{Tr}[\mathbf{Z}(\varepsilon)] &\leq \frac{N-1}{2} + (1-((1-\varepsilon)^{N-1})\sum_j q_j(\varepsilon)\mathbb{E}_{\boldsymbol{\mu}}[\tau_j(\varepsilon)] \\ &= \frac{N-1}{2} + (1-((1-\varepsilon)^{N-1})\mathrm{Tr}[\mathbf{Z}(\varepsilon)].\end{aligned}$$

It follows that

$$\begin{aligned}\mathrm{Tr}[\mathbf{Z}(\varepsilon)] &\leq \frac{N-1}{2}(1-\varepsilon)^{-(N-1)} \\ &= \frac{N-1}{2}\{1+(N-1)\varepsilon + O(\varepsilon^2)\} \\ &\leq \frac{N-1}{2} + \frac{(N-1)^2}{2}\varepsilon, \quad \text{if } \varepsilon < 1/N.\end{aligned}$$

□

Proposition 6.5 implies that if $\varepsilon < 1/N^2$, then for any reducible matrix $\mathbf{P}$, we have $\mathrm{Tr}[\mathbf{Z}(\varepsilon)] \geq N$, whereas for a Hamiltonian matrix $\mathbf{P}$, $\mathrm{Tr}[\mathbf{Z}(\varepsilon)]$ is not greater than $N/2$. Thus, if a Hamiltonian cycle exists, then for $\varepsilon < 1/N^2$ a reducible matrix cannot be a candidate for minimising $\mathrm{Tr}[\mathbf{Z}(\varepsilon)]$. The disadvantage of this approach, however, is that a very small perturbation does not resolve the aforementioned issues on computation and robustness of the objective function. Later in this section, we show that for the set of stochastic matrices, the condition $\varepsilon < 1/N^2$ can be relaxed to $\varepsilon < 1$.

The matrices with symmetric linear perturbation have recently received a lot of attention in computer science literature, mainly because these matrices are used in the Google PageRank algorithm that determines popularity of Web pages. The PageRank is defined as a stationary distribution of a Markov chain on a set of Web pages. This Markov chain serves as an elementary model of a surfing process. At each step, with probability $(1 - \varepsilon)$, a surfer follows a randomly chosen out-going hyperlink of a current page, and with probability ε, the surfer is bored and picks a new page on the Web at random. Clearly, a jump to a random page with probability ε is exactly the symmetric linear perturbation of a random walk on the Web graph, and the PageRank vector is the stationary vector of $\mathbf{P}^{\varepsilon}$. Google originally chose $\varepsilon = 0.15$. After the introduction of PageRank by Brin and Page [19], a lot of work has been done on analysing the formula and on determining the exact meaning of the parameter ε. We refer to Langville and Meyer [74] for an excellent survey of PageRank research. A number of explicit results are available on sensitivity of PageRank to small changes in the Web (see, for example, Bianchini *et al.* [11]), on the influence of the *damping factor* ε, and on the relation between the PageRank vector and the Web structure. However, a lion share of papers is devoted to enhancing the PageRank computation. A comprehensive overview on this topic can be found in Berkhin [10] and Langville and Meyer [74].

Langville and Meyer [74] and other PageRank literature show that as t tends to infinity, the powers $\mathbf{P}^k(\varepsilon)$ converge to $\mathbf{P}^*(\varepsilon)$ at least as fast as $(1 - \varepsilon)^k$ approaches zero. More specifically, Haveliwala and Kamvar [61] prove that the absolute value of the second largest eigenvalue (in modulus) of $\mathbf{P}^{\varepsilon}$ is at most $(1 - \varepsilon)$. Below we prove a stronger result for a doubly stochastic matrix $\mathbf{P}$ by stating the following simple linear algebraic fact.

The matrix $\mathbf{P}^{\varepsilon}$ is aperiodic; hence, the corresponding deviation matrix $\mathbf{Z}(\varepsilon)$ is defined by (6.2). Moreover, the results on the eigenvalues imply that the kth term of the series in (6.2) is of the order at most $(1 - \varepsilon)^k$, and thus the convergence improves significantly if ε is not very small. Clearly, numerical methods developed for computing PageRank can be applied for computing the deviation matrix $\mathbf{Z}(\varepsilon)$.

Assume that $\mathbf{P}$ is doubly stochastic. Then, the linearly perturbed matrix $\mathbf{P}^\varepsilon$ in (6.9) is also doubly stochastic, and we obtain an elementary relation

$$\mathbf{G}_{11}(\varepsilon) = \mathbf{Z}_{11}(\varepsilon) + q_1(\varepsilon) = \mathbf{Z}_{11}(\varepsilon) + 1/N. \tag{6.14}$$

Theorem 3.5 states that on the class of doubly stochastic matrices, $\mathbf{G}_{11}(\varepsilon)$ is minimised at the matrix inducing a Hamiltonian cycle for all ε in a small neighborhood of zero. The proofs in Borkar *et al.* [18] and in Chapter 3 rely heavily on analytic methods. In this section, we use a probabilistic approach to show that a Hamiltonian cycle ensures a minimal value of $[\mathbf{G}(\varepsilon)]_{11}$ for *any* $\varepsilon \in (0,1)$. This increases the potential applicability of the results in this line of research, because a reasonably large value of ε helps resolve some serious computational issues.

For notational convenience, we write

$$\mathbf{W}_*(\mathbf{P},\varepsilon) = \mathbf{I} - \mathbf{P}^\varepsilon + \mathbf{P}^*(\mathbf{P},\varepsilon), \tag{6.15}$$

and

$$\mathbf{W}(\mathbf{P},\varepsilon) = \mathbf{I} - \mathbf{P}^\varepsilon + 1/N\mathbf{J}. \tag{6.16}$$

The main result of the remainder of this chapter is given in the next theorem.

Theorem 6.2. *For a given Hamiltonian graph of order N and for $\varepsilon \in (0,1)$,*

$$\min_{\mathbf{P}} \mathrm{Tr}[\mathbf{G}(\mathbf{P},\varepsilon)] = \mathrm{Tr}[\mathbf{G}(\mathbf{P}_{\mathrm{H}},\varepsilon] = 1 + \frac{\varepsilon N - (1-(1-\varepsilon)^N)}{\varepsilon(1-(1-\varepsilon)^N)}, \tag{6.17}$$

for any $\mathbf{P}_{\mathrm{H}}$ corresponding to a Hamiltonian cycle.

The proof of Theorem 6.2 has the following structure: First, in Lemma 6.1, we establish the relationships between eigenvalues and eigenvectors of the matrix $\mathbf{P}$ and a particular matrix function of $\mathbf{P}$. Unless otherwise stated, the terms *eigenvectors* refer to right eigenvectors. The result of Lemma 6.1 gives rise to alternative formulas for the trace function, which we enunciate in Corollary 6.1. Next, in Lemma 6.2 we prove that the trace of the fundamental matrix $\mathbf{G}(\mathbf{P},\varepsilon)$ when $\varepsilon \in (0,1)$ for any randomised policy is bounded above by that of some deterministic policy, and bounded below by that of some other deterministic policy. This enables us to reduce our proof from the set of all stochastic policies to the set of all deterministic policies only. We derive the exact formulas for four exhaustive, mutually exclusive types of deterministic policies in Lemmas 6.3–6.6. Finally, we show that among these, Hamiltonian cycles are minimisers for the objective function.

Let $\mathbf{R}$ and $\mathbf{T}$ be arbitrary but fixed $N \times N$ rank-one stochastic matrices, that is, $\mathbf{R} = \boldsymbol{er}$ and $\mathbf{T} = \boldsymbol{ew}$ for some vectors $\boldsymbol{r}, \boldsymbol{w}$ such that $\boldsymbol{r}, \boldsymbol{w} \geq \mathbf{0}, \boldsymbol{re} = 1$ and $\boldsymbol{we} = 1$, where $\boldsymbol{e} = (1, \ldots, 1)^{\mathrm{T}}$. We define two matrices

$$\mathbf{P}_{\mathbf{R}}(\varepsilon) = (1 - \varepsilon)\mathbf{P} + \varepsilon\mathbf{R},$$

and

$$\mathbf{W}_{\mathbf{T}}(\mathbf{P}_{\mathbf{R}}(\varepsilon)) = \mathbf{I} - \mathbf{P}_{\mathbf{R}}(\varepsilon) + \mathbf{T}.$$

For $i = 1, \ldots, N$, we denote by λ_i the eigenvalues of $\mathbf{P}$, by η_i the eigenvalues of $\mathbf{P}_{\mathbf{R}}(\varepsilon)$, and by $\boldsymbol{s}$ the stationary vector of $\mathbf{W}_{\mathbf{T}}(\mathbf{P}_{\mathbf{R}}(\varepsilon))$, that is, $\boldsymbol{s}\mathbf{W}_{\mathbf{T}}(\mathbf{P}_{\mathbf{R}}(\varepsilon)) = \boldsymbol{s}$ and $\boldsymbol{se} = 1$. Thus, $\boldsymbol{s}\mathbf{P} = (1 - \varepsilon)^{-1}(\boldsymbol{w} - \varepsilon\boldsymbol{r})$.

The relationship between the eigenvalues λ_i's and the eigenvalues η_i's was given in Serra-Capizzano [96, Theorem 2.3]. Specifically,

$$\eta_i = \begin{cases} (1 - \varepsilon)\lambda_i & \text{for } i = 1, \ldots, N - 1, \\ \lambda_N = 1 & \text{for } i = N. \end{cases} \tag{6.18}$$

The following lemma gives the relationship between the eigenvalues and eigenvectors of $\mathbf{P}_{\mathbf{R}}(\varepsilon)$ and the eigenvalues and eigenvectors of $\mathbf{W}_{\mathbf{T}}(\mathbf{P}_{\mathbf{R}}(\varepsilon))$.

Lemma 6.1. *For $\varepsilon \in (0, 1)$ and any stochastic matrix $\mathbf{P}$, every eigenvector $\boldsymbol{v}_i$ of $\mathbf{P}_{\mathbf{Q}}(\varepsilon)$ corresponding to eigenvalue $\eta_i < 1$ gives an eigenvector $(\mathbf{I} - \boldsymbol{es})\boldsymbol{v}_i$ of $\mathbf{A}_{\mathbf{T}}(\mathbf{P}_{\mathbf{Q}}(\varepsilon))$ corresponding to eigenvalue $1 - \eta_i$. The eigenvector $\boldsymbol{e}$ of $\mathbf{P}_{\mathbf{Q}}(\varepsilon)$ corresponding to the eigenvalue 1 is an eigenvector of $\mathbf{A}_{\mathbf{T}}(\mathbf{P}_{\mathbf{Q}}(\varepsilon))$ corresponding to eigenvalue 1.*

Proof. Because $\mathbf{P}_{\mathbf{R}}(\varepsilon)$ is irreducible, the Perron-Frobenius theorem (see Horn and Johnson [66]) guarantees that its maximal eigenvalue 1 is simple and we can write its Jordan canonical form $\mathbf{L}$ as

$$\mathbf{L} = \begin{bmatrix} \mathbf{U} \\ \boldsymbol{\pi} \end{bmatrix} \mathbf{P}_{\mathbf{R}}(\varepsilon) \begin{bmatrix} \mathbf{V} \; \boldsymbol{e} \end{bmatrix}, \tag{6.19}$$

where $\mathbf{U}\boldsymbol{e} = \mathbf{0}, \boldsymbol{\pi}\mathbf{P}_{\mathbf{R}}(\varepsilon) = \boldsymbol{\pi}, \boldsymbol{\pi e} = 1, \boldsymbol{\pi}\mathbf{V} = \mathbf{0}, \mathbf{UV} = \mathbf{I}_{N-1}$ and $\mathbf{VU} = \mathbf{I}_N - \boldsymbol{e\pi}$. Then, it is straightforward that

$$\mathbf{L} = \begin{bmatrix} \mathbf{U}\mathbf{P}_{\mathbf{R}}(\varepsilon)\mathbf{V} & \mathbf{0} \\ \mathbf{0} & 1 \end{bmatrix}. \tag{6.20}$$

Let $\mathbf{C}$ be a Jordan canonical block with the same structure as that of $\mathbf{L}$, with eigenvalues $1 - \eta_i$ for $i = 1, \ldots, N - 1$ and the last eigenvalue being 1. We now show that $\mathbf{C}$ is the Jordan canonical form of $\mathbf{W}_{\mathbf{T}}(\mathbf{P}_{\mathbf{R}}(\varepsilon))$, that is,

$$\mathbf{C} = \begin{bmatrix} \mathbf{U} \\ \boldsymbol{s} \end{bmatrix} \mathbf{W}_{\mathbf{T}}(\mathbf{P}_{\mathbf{R}}(\varepsilon)) \begin{bmatrix} \mathbf{V} - \boldsymbol{es}\mathbf{V} \; \boldsymbol{e} \end{bmatrix}. \tag{6.21}$$

Expanding the right-hand side of (6.21) gives us

$$\begin{aligned}
&\begin{bmatrix} \mathbf{U} \\ \boldsymbol{s} \end{bmatrix} (\mathbf{I} - \mathbf{P_R}(\varepsilon) + \mathbf{T}) \begin{bmatrix} \mathbf{V} - \boldsymbol{es}\mathbf{V} \ \boldsymbol{e} \end{bmatrix} \\
&= \begin{bmatrix} \mathbf{U}(\mathbf{I} - \mathbf{P_R}(\varepsilon) + \mathbf{T})\mathbf{V} & \mathbf{0} \\ \mathrm{s}(\mathbf{V} - \boldsymbol{es}\mathbf{V}) & 1 \end{bmatrix} \\
&= \begin{bmatrix} \mathbf{I} - \mathbf{U}\mathbf{P_R}(\varepsilon)\mathbf{V} & \mathbf{0} \\ \mathrm{s}(\mathbf{I} - \mathbf{P_R}(\varepsilon) + \mathbf{T})(\mathbf{V} - \boldsymbol{es}\mathbf{V}) & 1 \end{bmatrix}.
\end{aligned}$$

Focusing on the bottom-left entry of the above right-hand side, we have

$$\begin{aligned}
\boldsymbol{s}(\mathbf{I} - \mathbf{P_R}(\varepsilon) + \mathbf{T})(\mathbf{V} - \boldsymbol{es}\mathbf{V}) &= \boldsymbol{s}(\mathbf{V} - (1-\varepsilon)\mathbf{P}\mathbf{V} - \varepsilon\boldsymbol{eq}\mathbf{V} + \boldsymbol{ew}\mathbf{V} - \boldsymbol{es}\mathbf{V}) \\
&= \boldsymbol{s}\mathbf{V} - (\boldsymbol{w} - \varepsilon\boldsymbol{r})\mathbf{V} - \varepsilon\boldsymbol{r}\mathbf{V} + \boldsymbol{w}\mathbf{V} - \boldsymbol{s}\mathbf{V} \\
&= 0,
\end{aligned}$$

which proves the lemma. □

Let $\boldsymbol{t}$ and $\boldsymbol{r}$ be the respective stationary vectors of $\mathbf{W}(\mathbf{P},\varepsilon)$ and $\mathbf{W}_*(\mathbf{P},\varepsilon)$, that is, $\boldsymbol{t}\mathbf{W}(\mathbf{P},\varepsilon) = \boldsymbol{t}$, $\boldsymbol{r}\mathbf{W}_*(\mathbf{P},\varepsilon) = \boldsymbol{r}$, and $\boldsymbol{te} = \boldsymbol{re} = 1$. We now easily obtain the eigenvalues and eigenvectors of the matrices $\mathbf{W}(\mathbf{P},\varepsilon)$, $\mathbf{W}_*(\mathbf{P},\varepsilon)$, and $\mathbf{G}(\mathbf{P},\varepsilon)$.

Corollary 6.1. *For $\varepsilon \in (0,1)$ and any stochastic matrix $\mathbf{P}$,*

(i) For $i = 1,\ldots,N-1$, every eigenvector $\boldsymbol{v}_i$ of $\mathbf{P}^\varepsilon$ corresponding to eigenvalue $\eta_i < 1$ gives an eigenvector $(\mathbf{I} - \boldsymbol{et})\boldsymbol{v}_i$ (respectively, $(\mathbf{I} - \boldsymbol{er})\boldsymbol{v}_i$) of $\mathbf{A}(\mathbf{P},\varepsilon)$ (respectively, $\mathbf{A}_*(\mathbf{P},\varepsilon)$) corresponding to eigenvalue $1 - \eta_i$. The eigenvector $\boldsymbol{e} = (1,\ldots,1)^{\mathrm{T}}$ of $\mathbf{P}$ corresponding to the unique eigenvalue 1 is an eigenvector of $\mathbf{A}(\mathbf{P},\varepsilon)$ (respectively, $\mathbf{A}_*(\mathbf{P},\varepsilon)$) corresponding to eigenvalue 1.

(ii) For $i = 1,\ldots,N-1$, the eigenvalues of $\mathbf{G}(\mathbf{P},\varepsilon)$ are 1 and $1/(1-(1-\varepsilon)\lambda_i)$.

(iii) $\mathrm{Tr}[\mathbf{G}(\mathbf{P},\varepsilon)] = 1 + \sum_{i=1}^{N-1} \dfrac{1}{1-(1-\varepsilon)\lambda_i} = \mathrm{Tr}[(\mathbf{I} - (1-\varepsilon)\mathbf{P})^{-1}] - (1-\varepsilon)/\varepsilon.$

Proof. Part (i) follows from Lemma 1 by setting $\mathbf{R} = 1/N\mathbf{J}$ and $\mathbf{T} = 1/N\mathbf{J}$ (respectively, $\mathbf{T} = \mathbf{P}^*(\mathbf{P},\varepsilon)$). We obtain Parts (ii) and (iii) by (6.18) and the well-known properties that the eigenvalues of an invertible matrix are the reciprocals of the eigenvalues of its inverse, and that the trace of a matrix is the sum of its eigenvalues. In the last equality of Part (iii), we note that the eigenvalues of $(\mathbf{I} - (1-\varepsilon)\mathbf{P})$ are ε and $1-(1-\varepsilon)\lambda_i$, for $i = 1,\ldots,N-1$. □

Lemma 6.1 indicates that the eigenvalues of $\mathbf{W_T}(\mathbf{P_R}(\varepsilon))$ do not depend on $\mathbf{T}$ and $\mathbf{R}$. This observation is crucial for our further discussion because instead of $\mathrm{Tr}[\mathbf{G}(\mathbf{P},\varepsilon)] = \mathrm{Tr}[\mathbf{A}_*^{-1}(\mathbf{P},\varepsilon)]$ we may consider $\mathrm{Tr}[\mathbf{A}^{-1}(\mathbf{P},\varepsilon)]$, which is beneficial as the elements of the matrix $\mathbf{A}(\mathbf{P},\varepsilon)$ depend linearly on the elements of $\mathbf{P}$. This allows us to make the next step in the proof of Theorem 6.2, by stating that the trace of the fundamental matrix can be maximised or minimised only at Markov chains associated with deterministic policies. This is formalized in the following lemma.

Lemma 6.2. *For any $\varepsilon \in (0,1)$ and for every randomised policy $\mathbf{P}$, there exist some deterministic policies $\mathbf{D}_1$ and $\mathbf{D}_2$ such that*

$$\mathrm{Tr}[\mathbf{G}(\mathbf{D}_1,\varepsilon)] \leq \mathrm{Tr}[\mathbf{G}(\mathbf{P},\varepsilon)] \leq \mathrm{Tr}[\mathbf{G}(\mathbf{D}_2,\varepsilon)]. \tag{6.22}$$

Proof. Let $\mathbf{P}$ be a randomised policy, and consider the randomisation at each row i of $\mathbf{P}$ separately. Suppose a particular row i is of the following structure:

$$[\cdots\ a\ \cdots\ b\ \cdots\ c\ \cdots], \tag{6.23}$$

where $a, b \in (0,1)$ and $a+b+c=1$. The subsequent arguments are easily generalisable to the case with more than three non-zero entries of the row in consideration. Consider a policy $\mathbf{P}_\gamma$ that coincides with $\mathbf{P}$ in all rows except row i, where it is replaced by

$$[\cdots\ \gamma\ \cdots\ \tfrac{(1-\gamma)}{1-a}b\ \cdots\ \tfrac{(1-\gamma)}{1-a}c\ \cdots], \quad \gamma \in [0,1]. \tag{6.24}$$

For $\gamma = a$, the matrix $\mathbf{P}_\gamma$ reduces to $\mathbf{P}$. By Corollary 6.1 and writing the inverse in terms of the adjoint,

$$\mathrm{Tr}[\mathbf{G}(\mathbf{P}_\gamma,\varepsilon)] = \mathrm{Tr}[\mathbf{W}^{-1}(\mathbf{P}_\gamma,\varepsilon)] = \sum_{i=1}^{N} \frac{\det \mathbf{W}_{ii}(\mathbf{P}_\gamma,\varepsilon)}{\det \mathbf{W}(\mathbf{P}_\gamma,\varepsilon)},$$

where $\mathbf{W}(\mathbf{P}_\gamma,\varepsilon) = \mathbf{I} - \mathbf{P}_\gamma^\epsilon + 1/N\mathbf{J}$ and $\mathbf{W}_{ii}(\mathbf{P}_\gamma,\varepsilon)$ is the matrix $\mathbf{W}(\mathbf{P}_\gamma,\varepsilon)$ with its ith row and ith column removed. As both $\det \mathbf{W}(\mathbf{P}_\gamma,\varepsilon)$ and $\det \mathbf{W}_{ii}(\mathbf{P}_\gamma,\varepsilon)$ are linear functions of γ for all $i = 1,\ldots,N$, we have

$$\mathrm{Tr}[\mathbf{G}(\mathbf{P}_\gamma,\varepsilon)] = \frac{C_1(\det \mathbf{W}(\mathbf{P}_\gamma,\varepsilon)) + C_2}{\det \mathbf{W}(\mathbf{P}_\gamma,\varepsilon)} = C_1 + \frac{C_2}{\det \mathbf{W}(\mathbf{P}_\gamma,\varepsilon)},$$

for some C_1, C_2 constant, $C_1 \neq 0$. Differentiating the objective function with respect to γ gives us

$$\frac{\partial}{\partial\gamma}\mathrm{Tr}\left[\mathbf{G}(\mathbf{P}_\gamma,\varepsilon)\right] = -\frac{C_3}{(\det \mathbf{W}(\mathbf{P}_\gamma,\varepsilon))^2},$$

for some constant C_3. Obviously, this partial derivative is zero for all $\gamma \in (0,1)$ if $C_3 = 0$, or never zero for all $\gamma \in (0,1)$ if $C_3 \neq 0$. In both cases, this implies that $\mathrm{Tr}[\mathbf{G}(\mathbf{P}_\gamma,\varepsilon)]$ is a monotone function over $\gamma \in [0,1]$, and is maximised or minimised at either extreme of the interval. As the ith row in $\mathbf{P}_{\gamma=0}$ or $\mathbf{P}_{\gamma=1}$ has at least one more zero than the ith row in $\mathbf{P}$, $\mathbf{P}_{\gamma=0}$ or $\mathbf{P}_{\gamma=1}$ has at least one more zero than $\mathbf{P}$, and

- either $\mathrm{Tr}[\mathbf{G}(\mathbf{P}_{\gamma=0},\varepsilon)]$ or $\mathrm{Tr}[\mathbf{G}(\mathbf{P}_{\gamma=1},\varepsilon)] \geq \mathrm{Tr}[\mathbf{G}(\mathbf{P},\varepsilon)]$, and
- either $\mathrm{Tr}[\mathbf{G}(\mathbf{P}_{\gamma=1},\varepsilon)]$ or $\mathrm{Tr}[\mathbf{G}(\mathbf{P}_{\gamma=0},\varepsilon)] \leq \mathrm{Tr}[\mathbf{G}(\mathbf{P},\varepsilon)]$, respectively.

Applying this process of increasing the number of zeros and consequently reducing the number of randomisations, we can find two deterministic policies $\mathbf{D}_1$ and $\mathbf{D}_2$ that satisfy the inequalities in (6.22). □

It follows from Lemma 6.2 that in order to prove Theorem 6.2, it suffices to compare $\mathrm{Tr}[\mathbf{G}(\mathbf{P}_{\mathrm{H}},\varepsilon)]$ to the values of $\mathrm{Tr}[\mathbf{G}(\mathbf{P},\varepsilon)]$, where $\mathbf{P}$ induces a deterministic policy. To this end, in Lemma 6.3 we derive the expression for $\mathrm{Tr}[\mathbf{G}(\mathbf{P}_{\mathrm{H}},\varepsilon)]$ and then, in Lemmas 6.4–6.6, we derive the expression for $\mathrm{Tr}[\mathbf{G}(\mathbf{P},\varepsilon)]$ corresponding to three other exhaustive and mutually exclusive classes of deterministic policies.

Lemma 6.3. *For any $\varepsilon \in (0,1)$ and any $\mathbf{P}_{\mathrm{H}}$ that corresponds to a Hamiltonian cycle, that is, a policy with a single ergodic class and no transient states,*

$$\mathrm{Tr}[\mathbf{G}(\mathbf{P}_{\mathrm{H}},\varepsilon)] = 1 + \frac{\varepsilon N - (1-(1-\varepsilon)^N)}{\varepsilon(1-(1-\varepsilon)^N)}.$$

Proof. Consider a Markov walk that starts at i and is governed by $\mathbf{P}$. The Markov walk governed by the strictly substochastic matrix $(1-\varepsilon)\mathbf{P}$ is a "stopped version" of this Markov walk in which at each time step there is a probability ε of termination. The termination time $T(\varepsilon)$ is geometrically-distributed with parameter ε. For $i \in \mathcal{S}$, the (i,i)th entry of $\{\mathbf{I}-(1-\varepsilon)\mathbf{P}\}^{-1}$ is equal to the expected number of visits to i in the time interval $[0,T(\varepsilon)]$, given that the random walk starts in state i. Now, assume that $\mathbf{P} = \mathbf{P}_{\mathrm{H}}$. Then, the random walk proceeds in cycles of length N, and thus, starting from any state i, the probability that it ever returns to i is $(1-\varepsilon)^N$, implying that the expected number of returns is $\{1-(1-\varepsilon)^N\}^{-1}$. Hence, from Part (iii) of Corollary 6.1 we obtain

$$\mathrm{Tr}[\mathbf{G}(\mathbf{P}_{\mathrm{H}},\varepsilon)] = \frac{N}{1-(1-\varepsilon)^N} - \frac{1-\varepsilon}{\varepsilon} = 1 + \frac{\varepsilon N - (1-(1-\varepsilon)^N)}{\varepsilon(1-(1-\varepsilon)^N)}.$$

□

Lemma 6.4. *For $\varepsilon \in (0,1)$ and for any $\mathbf{P}$ that corresponds to a deterministic policy with $\ell > 1$ ergodic classes and no transient states,*

$$\mathrm{Tr}[\mathbf{G}(\mathbf{P},\varepsilon)] = 1 + \frac{\ell - 1}{\varepsilon} + \sum_{i=1}^{\ell} \frac{m_i\varepsilon - (1-(1-\varepsilon)^{m_i})}{\varepsilon(1-(1-\varepsilon)^{m_i})}, \tag{6.25}$$

where m_i is the size of the ith ergodic class in $\mathbf{P}$.

Proof. Consider again the diagonal elements of $\{\mathbf{I}-(1-\varepsilon)\mathbf{P}\}^{-1}$. As $\mathbf{P}$ corresponds to a deterministic policy with $\ell > 1$ ergodic classes and no transient states, the Markov chain given by $\mathbf{P}$ contains ℓ disjoint cycles of lengths $m_1, \ldots, m_\ell$, with $\sum_{i=1}^{\ell} m_i = N$. For each of the cycles, we can apply the argument from the proof of Lemma 6.3. Then, every diagonal element that corresponds to a state in ergodic class i is given by $1/(1-(1-\varepsilon)^{m_i})$. Summing over all diagonal elements and using Corollary 6.1 (iii), we derive

$$\begin{aligned}\mathrm{Tr}[\mathbf{G}(\mathbf{P}_{\mathrm{H}}, \varepsilon)] &= \sum_{i=1}^{\ell} \frac{m_i}{1-(1-\varepsilon)^{m_i}} - \frac{1-\varepsilon}{\varepsilon} \\ &= 1 + \frac{\ell - 1}{\varepsilon} + \sum_{i=1}^{\ell} \frac{m_i \varepsilon - (1-(1-\varepsilon)^{m_i})}{\varepsilon(1-(1-\varepsilon)^{m_i})}.\end{aligned}$$

To obtain the last equation, it is sufficient to subtract and add $(\ell-1)(1-\varepsilon)/\varepsilon$ in the second expression and then use the result of Lemma 6.3. □

Lemma 6.5. *For any $\varepsilon \in (0,1)$ and any $\mathbf{P}$ that corresponds to a policy with a single ergodic class and one or more transient states,*

$$\mathrm{Tr}[\mathbf{G}(\mathbf{P}, \varepsilon)] = (N - m + 1) + \frac{m\varepsilon - (1-(1-\varepsilon)^m)}{\varepsilon(1-(1-\varepsilon)^m)},$$

where $m < N$ is the size of the single ergodic class.

Proof. Let i be a transient state of the Markov chain with transition matrix $(1-\varepsilon)\mathbf{P}$. Since $\mathbf{P}$ is deterministic, a Markov random walk starting at i can never return to i. Since the (i,i)th element of $\{\mathbf{I}-(1-\varepsilon)\mathbf{P}\}^{-1}$ is the average number of visits to i on $[0, T(\varepsilon)]$ starting from i, where $T(\varepsilon)$ is defined as in the proof of Lemma 6.3, we conclude that each transient state contributes precisely one to $\mathrm{Tr}[[\mathbf{I}-(1-\varepsilon)\mathbf{P}]^{-1}]$. On the other hand, ergodic states form a cycle of length m, and we can determine the contribution of these states by applying the argument as in the proof of Lemma 6.3 with $N = m$. Summing the contributions of transient and ergodic states and applying Corollary 6.1 (iii), we obtain the result of the lemma. □

Lemma 6.6. *For any $\varepsilon \in (0,1)$ and for any $\mathbf{P}$ corresponding to a policy with multiple ergodic classes and one or more transient states,*

$$\mathrm{Tr}[\mathbf{G}(\mathbf{P}, \varepsilon)] = N - \sum_{i=1}^{\ell} m_i + 1 + \frac{\ell - 1}{\varepsilon} + \sum_{i=1}^{\ell} \frac{m_i \varepsilon - (1-(1-\varepsilon)^{m_i})}{\varepsilon(1-(1-\varepsilon)^{m_i})}, \quad (6.26)$$

where m_i is the size of the ith ergodic class in $\mathbf{P}$.

Proof. The result follows by combining the proofs of Lemmas 6.4 and 6.5. □

Now we are ready to prove Theorem 6.2.

Proof of Theorem 6.2. We need to show that for any $\varepsilon \in (0,1)$ and for any stochastic policy $\mathbf{P}$ feasible on a given Hamiltonian graph, Hamiltonian cycles are indeed the minimisers.

As the result of Lemma 6.2 enables us to reduce the proof for the set of stochastic policies to the proof for the set of deterministic policies, by Lemmas 6.3–6.6 all we need to show now is, that for $\ell > 1$ and $m, m_i < N$ with $\sum_{i=1}^{\ell} m_i \leq N$,

$$1 + \frac{\varepsilon N - (1-(1-\varepsilon)^N)}{\varepsilon(1-(1-\varepsilon)^N)} \leq 1 + \frac{\ell - 1}{\varepsilon} + \sum_{i=1}^{\ell} \frac{m_i\varepsilon - (1-(1-\varepsilon)^{m_i})}{\varepsilon(1-(1-\varepsilon)^{m_i})}, \tag{6.27}$$

$$1 + \frac{\varepsilon N - (1-(1-\varepsilon)^N)}{\varepsilon(1-(1-\varepsilon)^N)} \leq (N - m + 1) + \frac{m\varepsilon - (1-(1-\varepsilon)^{m})}{\varepsilon(1-(1-\varepsilon)^{m})}, \tag{6.28}$$

and

$$1 + \frac{\varepsilon N - (1-(1-\varepsilon)^N)}{\varepsilon(1-(1-\varepsilon)^N)}$$
$$\leq N - \sum_{i=1}^{\ell} m_i + 1 + \frac{\ell - 1}{\varepsilon} + \sum_{i=1}^{\ell} \frac{m_i\varepsilon - (1-(1-\varepsilon)^{m_i})}{\varepsilon(1-(1-\varepsilon)^{m_i})}. \tag{6.29}$$

Note that for m_i such that $\sum_{i=1}^{\ell} m_i = N$, inequality (6.29) reduces to inequality (6.27). Therefore, we only need to focus on the last two inequalities (6.28) and (6.29). In order to prove (6.28), it suffices to show that

$$\frac{N}{1-(1-\varepsilon)^N} \leq N - m + \frac{m}{1-(1-\varepsilon)^m}. \tag{6.30}$$

Define $t = 1 - \varepsilon$, then $t \in (0, 1)$. Writing both sides over a common denominator, inequality (6.30) is equivalent to

$$\frac{Nt^N}{1-t^N} \leq \frac{mt^m}{1-t^m}. \tag{6.31}$$

Define the function $g(x) = xt^x/(1-t^x)$ for $x \geq 0$. Differentiating $g(x)$ gives us

$$\frac{d}{dx} g(x) = \frac{t^x(1 - t^x + x \ln t)}{(1-t^x)^2}.$$

It is straightforward that the denominator $(1-t^x)^2 > 0$ for all $x > 0$. Also, $t^x > 0$ for all $x \geq 0$. Define the function $h(x) = 1 - t^x + x \ln t$ for $x \geq 0$, and observe that $h(0) = 0$ and

$$\frac{d}{dx} h(x) = -t^x \ln t + \ln t = \ln t(1 - t^x) < 0.$$

as $\ln t < 0$ and $1 - t^x > 0$, for all $x > 0$. Therefore, $h(x)$ is a decreasing function and $h(x) < 0$ for $x > 0$. This implies that $\frac{d}{dx}g(x) < 0$ for $x > 0$ and consequently $g(x)$ is a decreasing function. As $N > m$, we obtain (6.31) and subsequently (6.28).

Now, to prove inequality (6.29), it suffices to show that

$$\frac{N}{1-(1-\varepsilon)^N} \leq N - \sum_{i=1}^{\ell} m_i + \frac{\ell - 1}{\varepsilon} + \sum_{i=1}^{\ell}\{\frac{m_i}{1-(1-\varepsilon)^{m_i}} - \frac{1}{\varepsilon}\}, \tag{6.32}$$

which is equivalent to

$$\frac{N(1-\varepsilon)^N}{1-(1-\varepsilon)^N} + \frac{1}{\varepsilon} \leq \frac{m_1(1-\varepsilon)^{m_1}}{1-(1-\varepsilon)^{m_1}} + \sum_{i=2}^{\ell} \frac{m_i(1-\varepsilon)^{m_i}}{1-(1-\varepsilon)^{m_i}}. \tag{6.33}$$

As we have proved that $g(x)$ is a decreasing function for $x > 0$,

$$\frac{N(1-\varepsilon)^N}{1-(1-\varepsilon)^N} \leq \frac{m_1(1-\varepsilon)^{m_1}}{1-(1-\varepsilon)^{m_1}}.$$

To obtain (6.33), we require

$$\frac{1}{\varepsilon} \leq \frac{m_i(1-\varepsilon)^{m_i}}{1-(1-\varepsilon)^{m_i}}, \tag{6.34}$$

or equivalently,

$$0 \leq \varepsilon m_i(1-\varepsilon)^{m_i} - (1-(1-\varepsilon)^{m_i}) = (1+\varepsilon m_i)(1-\varepsilon)^{m_i} - 1. \tag{6.35}$$

for $i = 2, \ldots, \ell$. Define $f_i(x) = (1 + x m_i)(1-x)^{m_i} - 1$, for $i = 1, \ldots, \ell$ and $x \in [0, 1)$. We observe that $f_i(0) = 0$, and differentiating $f_i(x)$ gives

$$\begin{aligned}\frac{d}{dx}f(x) &= m_i(1-x)^{m_i} - m_i(1+xm_i)(1-x)^{m_i-1} \\ &= m_i(1-x)^{m_i-1}\{(1-x) - (1+xm_i)\} \\ &= xm_i(1-x)^{m_i-1}(m_i - 1) > 0\end{aligned}$$

for all $x \in (0, 1)$, as $2 \leq m_i \leq N - 2$. This implies that $f_i(x)$ is an increasing function, and $f_i(x) > 0$ for $x \in (0, 1)$. Thus, we obtain (6.35) and subsequently (6.29). □

In summary, in this chapter, we concluded our exploration of functionals based on Markov chains (induced by stationary policies) by focussing on the trace of fundamental matrices corresponding to such policies. We demonstrated that in both the perturbed and unperturbed cases these trace functionals can be used as objective functions of minimisation problems that achieve their global minima only at policies identifying Hamiltonian cycles, whenever the underlying graph is Hamiltonian. We note that, to date, there has been no attempt to exploit these results to develop any algorithmic methods to find Hamiltonian cycles.

Part IV
Algorithms

Chapter 7
Linear Programming Based Algorithms

7.1 Introduction

In Chapter 4, we showed that when a graph is embedded in a suitably constructed Markov decision process, the associated convex domain of discounted occupational measures is a polyhedron with extreme points corresponding to all spanning subgraphs of the given graph. Furthermore, from Theorem 4.1 we learned that a simple cut of the above domain yields a polyhedron the extreme points of which correspond to only two possible types: Hamiltonian cycles and convex combinations of short and noose cycles. These properties, naturally, suggest certain algorithmic approaches to searching for Hamiltonian cycles.

In this chapter, we present two methods to solve the HCP: the *branch and fix method*, and the *Wedged-MIP heuristic*. Both methods take advantage of the Markov decision process embedding outlined in Chapter 4. The branch and fix method is implemented in MATLAB and results that are supplied demonstrate the potential of this model. The Wedged-MIP heuristic is implemented in IBM ILOG OPL-CPLEX and succeeds in solving large graphs, including two of the large test problems given on the TSPLIB website maintained by the University of Heidelberg [98]. Both of these methods operate in the space of discounted occupational measures, but similar methods could be developed for the space of limiting average occupational measures.

Here, we continue exploiting properties of the space of discounted occupational measures in the Markov decision process associated with a graph G, as outlined in Chapter 4. In particular, we apply the non-standard branch and bound method of Filar and Lasserre [50] to Feinberg's embedding of the HCP in a discounted Markov decision process [44] (rather than the limiting average Markov decision process used previously). This embedding has the benefit that the discount parameter does not destroy the sparsity of coefficient matrices to nearly the same extent as did the perturbation parameter ε, used in [50] to replace the underlying probability transitions $p(j|i,a)$ of the MDP by the linearly perturbed transitions $p^{\varepsilon}(j|i,a)$. We refer to the method that arises from this embedding as the branch and fix method[1].

We show that in the present application, the appropriate sub-space of discounted occupational measures is synonymous with a polytope $\widetilde{\mathcal{X}}_\beta$ defined by only $N+1$ equality constraints and nonnegativity constraints. Using results in Theorems 4.1 and 4.2 about the structure of extreme points of $\widetilde{\mathcal{X}}_\beta$, we conjecture that Hamiltonian cycles will be found far earlier, and the resulting *logical branch and fix tree* will have fewer branches than that for more common polytopes. The logical branch and fix (B&F) tree that arises from the branch and fix method is a rooted tree. The root of the logical B&F tree corresponds to the original graph G, and each branch corresponds to a certain fixing of arcs in G. Then, a branch forms a pathway from the root of the logical B&F tree to a leaf. These leaves correspond to particular subdigraphs of G, which may or may not contain Hamiltonian cycles. At the maximum depth of the logical B&F tree, each leaf corresponds to a subdigraph for which there is exactly one arc emanating from every vertex. We refer to subdigraphs of this type as *spanning 1-out-regular subdigraphs* of G. Leaves at a shallower level correspond to subdigraphs in which there are multiple arcs emanating from at least one vertex.

The set of all spanning 1-out-regular subdigraphs has a one-to-one correspondence with the set of all deterministic policies in G. Even for graphs with bounded out-degree, this represents a set with non-polynomial cardinality. Cubic graphs, for example, have 3^N distinct deterministic policies. Hence, it is desirable to be able to fathom branches early, and consequently restrict the number of leaves in the logical B&F tree. The special structure of the extreme points of $\widetilde{\mathcal{X}}_\beta$ usually enables us to identify a Hamiltonian cycle before obtaining a spanning 1-out-regular subdigraph, limiting the depth of the logical B&F tree. We achieve significant improvements by introducing into the branch and fix method additional feasibility constraints as bounds, and logical checks that allow us to fathom branches early. This further limits the depth of the logical B&F tree.

The resulting method is guaranteed to solve the HCP in finitely many iterations. While the worst case may involve examination of exponentially many branches, empirically we show that the number of branches required to find a Hamiltonian cycle is generally reduced to a tiny fraction of the total number of deterministic policies. For example, a 24-vertex Hamiltonian cubic graph has $3^{24} \approx 3 \times 10^{11}$ possible choices for deterministic policies, but the algo-

---

[1] Since the speed of convergence depends more on arc fixing features than on bounds, the name *branch and fix* (or B&F) method is more appropriate than *branch and bound.*

rithm finds a Hamiltonian cycle by examining only 28 branches. We observe that Hamiltonian graphs perform better than non-Hamiltonian graphs, as they typically have many Hamiltonian cycles spread throughout the logical B&F tree, and only one needs to be found. However, even in non-Hamiltonian graphs we demonstrate that the algorithm performs well. For instance, a 28-vertex non-Hamiltonian cubic graph has $3^{28} \approx 2 \times 10^{13}$ possible choices for deterministic policies, but the algorithm terminates after investigating only 11,708 branches. This example highlights the ability of the B&F method to fathom branches early, allowing us to ignore, in this case, 99.99999995% of the potential branches.

In addition to the basic branch and fix method, we develop and compare several branching methods for traversing the logical B&F tree that may find Hamiltonian cycles quicker in certain graphs, and propose additional constraints that can find infeasibility at an earlier depth in the logical B&F tree. We provide experimental results demonstrating the significant improvement achieved by these additions. We also demonstrate that $\widetilde{\mathcal{X}}_\beta$ can be a useful polytope in many other optimisation algorithms. In particular we use $\widetilde{\mathcal{X}}_\beta$, along with the additional constraints, in a mixed integer programming model that can solve extremely large graphs using commercially available software such as CPLEX. Finally, we present solutions of four large non-regular graphs, with 250, 500, 1000 and 2000 vertices respectively, which are obtained by this model.

7.2 Branch and Fix Method

In this section, we describe the branch and fix method and some of the techniques used within that branch and fix method that are designed to help limit the size of the logical branching tree. The original source for this algorithm is Ejov *et al.* [31] and the presentation here is based on the PhD thesis of Haythorpe [62].

Outline of The Branch and Fix Method

In view of the fact that it is only 1-randomised policies that prevent standard simplex methods from finding a Hamiltonian cycle, it has been recognised for some time that branch and bound-type methods can be used to eliminate the possibility of arriving at these undesirable extreme points (see, for instance, Filar and Lasserre [50]). However, the method reported in Filar and Lasserre [50] uses an embedding in a long-run average MDP, with a perturbation of transition probabilities that introduces a small parameter in most of the $p(j|i, a)$ coefficients of variables in linear constraints (4.18), thereby leading to loss of sparsity. Furthermore, the method in Filar and Lasserre [50] was never implemented fully, or tested beyond a few simple examples.

Theorem 4.1 indicates that 1-randomised policies induced by extreme points of $\widetilde{\mathcal{X}}_\beta$ are less prevalent than might have been conjectured, since they cannot be constructed from convex combinations of just any two deterministic policies. This provides motivation for testing algorithmic approaches based on successive elimination of arcs that could be used to construct these convex combinations. Since our goal is to find an extreme point $\boldsymbol{x}_e \in \widetilde{\mathcal{X}}_\beta$ such that

$$\boldsymbol{f} = M^{-1}(\boldsymbol{x}_e) \in \mathcal{F}_\mathcal{D},$$

we have a number of degrees of freedom in designing an algorithm. In particular, different linear objective functions can be chosen at each stage of the algorithm, the parameter $\beta \in (0,1)$ can be adjusted, and $\mu \in (0, 1/N)$ can be chosen small but not so small as to cause numerical difficulties. The latter parameter needs to be positive to ensure that the inverse map M^{-1} is well-defined. In the experiments reported here, we choose μ to be $1/N^2$.

The branch and fix method is as follows. We solve a sequence of linear programs—two at each branching point of the logical B&F tree—with the generic structure

$$\begin{aligned} &\min L(\boldsymbol{x}) \\ \text{subject to} \quad & \\ &\boldsymbol{x} \in \widetilde{\mathcal{X}}_\beta, \text{and additional constraints, if any, on arcs fixed earlier.} \end{aligned} \tag{7.1}$$

Step 1—Initiation. We solve the original LP (7.1) without any additional constraints and with some choice of an objective function $L(\boldsymbol{x})$, to obtain an optimal basic feasible solution $\boldsymbol{x}_0$. We then find $\boldsymbol{f}_0 = M^{-1}(\boldsymbol{x}_0)$. If $\boldsymbol{f}_0 \in \mathcal{F}_\mathcal{D}$, we stop, the policy $\boldsymbol{f}_0$ identifies a Hamiltonian cycle. Otherwise, $\boldsymbol{f}_0$ is a 1-randomised policy.

Step 2—Branching. We use the 1-randomised policy $\boldsymbol{f}_0$ to identify the splitting vertex i, and two arcs (i, j_1) and (i, j_2) corresponding to the single randomisation in $\boldsymbol{f}_0$. If there are d arcs $\{(i, a_1), \ldots, (i, a_d)\}$ emanating from vertex i, we construct d subdigraphs: $G_1, G_2, \ldots, G_d$, where in G_k the arc (i, a_k) is the only arc emanating from vertex i. These graphs are identical to the original graph G at all other vertices. In this process we, by default, fix an arc in each G_k.

Step 3—Fixing. In many subdigraphs, the fixing of one arc implies that other arcs may also be fixed[2], without a possibility of unintentionally elim-

[2] This frequently happens in the case of cubic graphs that supplied many of our test examples. For instance, see Figure 7.2.

inating a Hamiltonian cycle containing already fixed arcs that are part of a Hamiltonian cycle in the current subdigraph. Later in this section, we describe four checks for determining additional arcs that can be fixed. Once we identify these arcs, we also fix them at this step.

Step 4—Iteration. We solve a second LP (with the objective function (7.9)) to determine if (4.19) is still satisfied with the current fixing of arcs. If so, we repeat Step 1 with the LP (7.1) constructed for the graph at the current branching point of the logical B&F tree, with additional constraints derived in (7.5) and (7.6) below. This branching point may correspond to $G_1, G_2, \dots, G_d$, or to a sub-graph constructed from one of these with the help of additional arc fixing[3].

We now briefly discuss the construction of additional constraints alluded to in Step 4 of the B&F. If $\boldsymbol{f}$ is a Hamiltonian policy, $\boldsymbol{x} = M(\boldsymbol{f})$, and $\mu = 0$, then we can easily check that $\boldsymbol{x}$ satisfies (4.18)–(4.20) and, for $k = 0, \dots, N-1$,

$$x_{i_k i_{k+1}} = \sum_{a \in \mathcal{A}(i_k)} x_{i_k a} = \frac{\beta^k}{1-\beta^N} \tag{7.2}$$

where (i_k, i_{k+1}) is the kth arc on the Hamiltonian cycle traced out by $\boldsymbol{f}$. This immediately suggests lower and upper bounds on sums of the $\boldsymbol{x}$ variables corresponding to arcs emanating from the heads of fixed arcs. This is because if $i_{k+1} \neq 1$,

$$\sum_{a \in \mathcal{A}(i_{k+1})} x_{i_{k+1} a} - \beta x_{i_k i_{k+1}} = 0. \tag{7.3}$$

If $i_{k+1} = 1$, then $k+1 = N$ and we have

$$-\beta^N \sum_{a \in \mathcal{A}(1)} x_{1a} + \beta x_{i_{N-1},1} = 0. \tag{7.4}$$

For $\mu > 0$, analogous, but more complex, expressions for the preceding sums can be derived and the relationship (7.3) between these sums at successive vertices on the Hamiltonian cycle, for $i_{k+1} \neq 1$, is simply

$$\sum_{a \in \mathcal{A}(i_{k+1})} x_{i_{k+1} a} - \beta x_{i_k i_{k+1}} = \mu. \tag{7.5}$$

If the fixed arc is the final arc $(i_N, 1)$, we have

[3] As is typical with branching methods, decisions guiding which branch to select first are important and open to alternative heuristics. We investigate five possible branching methods later in this section.

$$-\beta^N \sum_{a \in \mathcal{A}(1)} x_{1a} + \beta x_{i_{N-1},1} = \frac{\mu\beta(1-\beta^{N-1})}{1-\beta}. \tag{7.6}$$

We derive equation (7.5) by simply inspecting the form of (4.18). For (7.6), we know from (4.18) that

$$\sum_{a \in \mathcal{A}(1)} x_{1a} - \beta x_{i_{N-1},1} = 1 - (N-1)\mu,$$

and therefore

$$\beta x_{i_{N-1},1} = \sum_{a \in \mathcal{A}(1)} x_{1a} - 1 + (N-1)\mu. \tag{7.7}$$

Then, we substitute (7.7) into the left-hand side of (7.6) to obtain

$$-\beta^N \sum_{a \in \mathcal{A}(1)} x_{1a} + \beta x_{i_{N-1},1} = (1-\beta^N) \sum_{a \in \mathcal{A}(1)} x_{1a} - 1 + (N-1)\mu. \tag{7.8}$$

Finally, we substitute (4.19) into (7.8) to obtain

$$\begin{aligned} &- \beta^N \sum_{a \in \mathcal{A}(1)} x_{1a} + \beta x_{i_{N-1},1} \\ &= (1-\beta^N)\frac{(1-(N-1)\mu)(1-\beta) + \mu(\beta-\beta^N)}{(1-\beta)(1-\beta^N)} - 1 + (N-1)\mu \\ &= \frac{\mu\beta(1-\beta^{N-1})}{1-\beta}, \end{aligned}$$

which coincides with (7.6).

Structure of the underlying LP in the branch and fix method

At the initiation step of the B&F method, we solve a feasibility problem of satisfying constraints (4.18)–(4.20). This allows us to determine on which vertex to begin branching.

At every branching point of the logical B&F tree other than the root, we solve an additional LP that attempts to determine if we need to continue exploring the current branch. As the algorithm evolves along successive branching points of the logical B&F tree, we have additional information about which arcs have been fixed. This permits us to perform tests to check the possibility of finding a Hamiltonian cycle containing these fixed arcs. If we determine that it is impossible, we fathom that branching point of the logical B&F tree and no further exploration of that branch is required. For instance, suppose that all fixed arcs belong to a set $\mathcal{U}$. Let the objective function of a second LP[4] be

$$L(\boldsymbol{x}) = \sum_{a\in\mathcal{A}(1)} x_{1a}, \tag{7.9}$$

and minimise (7.9) subject to constraints (4.18) and (4.20) together with equations (7.5) and (7.6) providing additional constraints for each arc in $\mathcal{U}$. If the minimum $L^*(\boldsymbol{x})$ fails to reach the level defined by the right-hand side of the now omitted constraint (4.19) of $\widetilde{\mathcal{X}}_\beta$, or if the constraints are infeasible, then there exists no Hamiltonian cycle that uses all the arcs of $\mathcal{U}$, and we fathom the current branching point of the logical B&F tree. Otherwise, we solve the LP (7.1) with the objective function[5]

$$L(\boldsymbol{x}) = \sum_{(i,j)\in\mathcal{U}} \{ \sum_{a\in\mathcal{A}(j)} x_{ja} - \beta \sum_{a\in\mathcal{A}(i)} x_{ia} \}, \tag{7.10}$$

and with no additional constraints beyond those in $\widetilde{\mathcal{X}}_\beta$. This LP will either find a Hamiltonian cycle, or it will lead to an extreme point $\boldsymbol{x}'_e$ such that $\boldsymbol{f}' = M^{-1}(\boldsymbol{x}'_e)$ is a new 1-randomised policy. Of course, alternative objective functions $L(\boldsymbol{x})$ could also be considered.

Arc fixing checks

There are a number of logical checks that enable us to fix additional arcs once a decision is taken to fix a particular arc. This is best illustrated with the help of an example. These checks are in the spirit of well-known rules for constructing Hamiltonian cycles (see Tucker [99, Section 8.2]).

Consider the envelope graph (Figure 7.1). Figure 7.2 shows the kind of logical additional arc fixing that can arise.

Check 1: Consider the top-left graph in Figure 7.2. The fixed arcs are $(1,2)$ and $(6,3)$. Since the only arcs that can go to vertex 5 are $(1,5)$, $(4,5)$ and $(6,5)$, we may also fix arc $(4,5)$ as vertices 1 and 6 already have fixed arcs going elsewhere. In this case, we say that arc $(4,5)$ is *free*, whereas arcs $(1,5)$ and $(6,5)$ are not free. In general, if only one free arc enters a vertex, it must be fixed.

4 Although we call (7.9) the second LP, it is the first LP solved in all iterations other than the initial iteration. Since it is not solved first in the initial iteration, we refer to (7.9) as the second LP.

5 For simplicity, we are assuming here that $\mathcal{U}$ does not contain any arc going into vertex 1. If such an arc were in $\mathcal{U}$, the objective function (7.10) would have one term consistent with the left-hand side of equation (7.6).

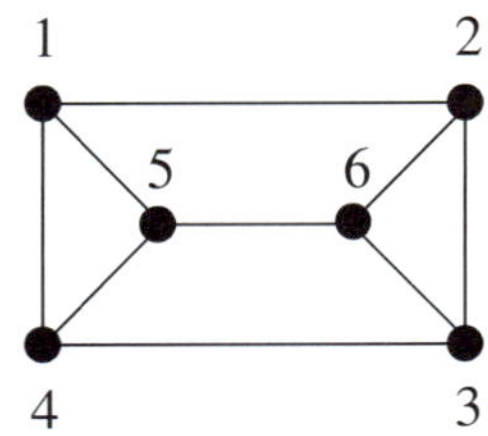

Fig. 7.1: The envelope graph

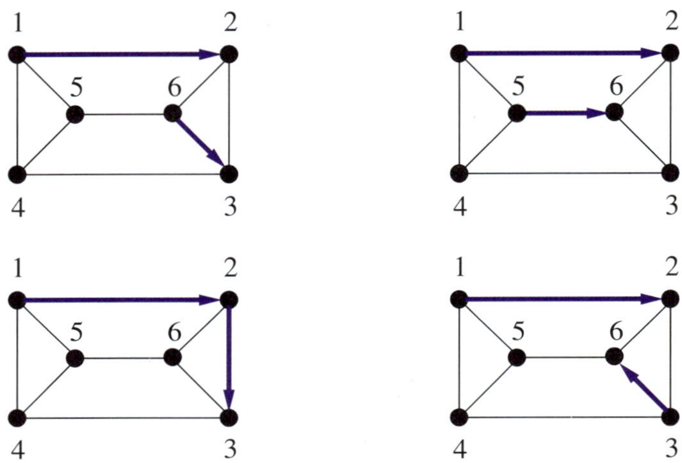

Fig. 7.2: Various arc fixing situations

Check 2: Consider the top-right graph in Figure 7.2. The fixed arcs are $(1, 2)$ and $(5, 6)$. The arcs going to vertex 5 are $(1, 5)$, $(4, 5)$ and $(6, 5)$. We cannot choose $(6, 5)$ as this would create a subcycle of length 2, and vertex 1 already has a fixed arc going elsewhere, so we must fix arc $(4, 5)$. In general, if there are only two free arcs and fixing one arc would create a subcycle, we must fix the other one.

Check 3: Consider the bottom-left graph in Figure 7.2. The fixed arcs are $(1, 2)$ and $(2, 3)$. Since the only arcs that can come from vertex 6 are $(6, 2)$, $(6, 3)$ and $(6, 5)$, we must fix arc $(6, 5)$ as vertices 2 and 3 already have arcs going into them. In this case, we say that arcs $(6, 3)$ and $(6, 5)$ are *blocked*, whereas arc $(6, 2)$ is *unblocked*. In general, if there is only one unblocked arc emanating from a vertex, the arc must be fixed.

Check 4: Consider the bottom-right graph in Figure 7.2. The fixed arcs are $(1, 2)$ and $(3, 6)$. The only arcs that can come from vertex 6 are $(6, 2)$, $(6, 3)$ and $(6, 5)$. We cannot choose $(6, 2)$ because vertex 2 already has an incoming arc, and we cannot choose $(6,3)$ as this would create a subcycle, so we must fix arc $(6,5)$. In general, if there are two unblocked arcs emanating from a vertex and fixing one arc would create a subcycle, we must fix the other one.

The branch and bound method given in Filar and Lasserre [50] always finds a Hamiltonian cycle if one exists. While the branch and fix method presented here is in the same spirit as the method in [50], we include a finite convergence proof for the sake of completeness.

Theorem 7.1. *The branch and fix method converges in finitely many steps. In particular, if G is Hamiltonian, the algorithm finds a Hamiltonian cycle; otherwise, the algorithm terminates after fathoming all constructed branches of the logical B&F tree.*

Proof. Given a graph G, at each stage of the algorithm, a splitting vertex is identified and branches are created for all arcs emanating from that vertex. As we consider every arc for this vertex and therefore explore every possibility from this vertex, the branching process cannot prevent the discovery of a Hamiltonian cycle if one exists in the graph G. It then suffices to confirm that none of the checking, bounding, or fixing steps in the branch and fix method can eliminate the possibility of finding a Hamiltonian cycle.

Recall that constraints (4.19), (7.5) and (7.6) are shown to be satisfied by all Hamiltonian cycles. Then, for a particular branching point, if the minimum value of (7.9) constrained by (4.18), (4.20), (7.5) and (7.6) cannot achieve the value given in (4.19), or if the constraints are infeasible, there cannot be any Hamiltonian cycles remaining in the subdigraph. Therefore, fathoming the branch due to the second LP (with the objective function (7.9)) cannot eliminate any Hamiltonian cycles. Checks 1–4 above are designed to ensure that at least one arc goes into and comes out of each vertex (while preventing the formation of subcycles), by fixing one or more arcs in situations where any other choice would violate this requirement. Since this is a requirement for all Hamiltonian cycles, it follows that the arc fixing procedures previously described cannot eliminate any Hamiltonian cycles.

The B&F method continues to search the tree until either a Hamiltonian cycle is found, or all constructed branches are fathomed. Since none of the steps in the B&F method can eliminate the possibility of finding a Hamiltonian cycle, we are guaranteed to find one of the Hamiltonian cycles in G. If all branches of the logical B&F tree are fathomed without finding any Hamiltonian cycles, we can conclude that G is non-Hamiltonian. □

While the branch and fix method only finds a single Hamiltonian cycle, it is possible to find all Hamiltonian cycles by simply recording each Hamiltonian cycle when it is found, and then continuing to search the branch and fix tree rather than terminating.

Corollary 7.1. *The depth of the logical B&F tree has an upper bound of N.*

Proof. Since at each branching point of the B&F we branch on all arcs emanating from a vertex, it follows that once an arc (i, j) is fixed, no other arcs emanating from vertex i can be fixed. Then, at each level of the branch and fix tree, a different vertex is branched on. After N levels, all vertices will have exactly one arc fixed, and either a Hamiltonian cycle will be found, or the relevant LP will be infeasible and we will fathom that branch. □

In practice, the arc fixing checks ensure that we never reach this upper bound, as we will certainly fix multiple arcs at branching points corresponding to subdigraphs where only few unfixed arcs remain.

7.3 An Algorithm that Implements the Branch and Fix Method

In Section 7.2, we described the branch and fix method for the HCP and proved its finite convergence. Here, we present a recursive algorithm that implements the method in pseudocode format, with separate component algorithms for the arc fixing checks and for solving the second LP. The input variable *fixed arcs* is initially input as an empty vector, as no arcs are fixed at the commencement of the algorithm. The output term *HC* may either be a Hamiltonian cycle found by the branch and fix method, or a message that no Hamiltonian cycle was found.

```
Input: G, fixed arcs
Output: G, fixed arcs

begin
        N ← Size(G)
        for i from 1 to N
            if only one arc (j, i) is free to go into vertex i
                  fixed arcs ← Add arc (j, i) to fixed arcs if it is not already in fixed arcs
            end

            if two arcs (j, i), (k, i), j ≠ k are free to go into i and arc (i, k) is in fixed arcs
                  fixed arcs ← Add arc (j, i) to fixed arcs if it is not already in fixed arcs
            end

            if only one arc (i, j) that emanates from i is unblocked
                  fixed arcs ← Add arc (i, j) to fixed arcs if it is not already in fixed arcs
            end

            if two arcs (i, j), (i, k), j ≠ k that emanate from i are unblocked, and arc (k, i)
            is in fixed arcs
                  fixed arcs ← Add arc (i, j) to fixed arcs if it is not already in fixed arcs
            end
        end
end
```

Algorithm 7.1: Checking algorithm

```
Input: G, β, fixed arcs
Output: HC

begin
        N ← Size(G)
        μ ← 1/N^2
        function value ← Algorithm 7.3: Second LP algorithm(G, β, fixed arcs)
        if infeasibility is found or function value > ((1 − (N − 1)μ)(1 − β) + μ(β − β^N)) / ((1 − β)(1 − β^N))
            return no HC found
        end
        X̃_β ← constraints (4.18)–(4.20)
        for Each arc in fixed arcs
            if Arc goes into vertex 1
                X̃_β ← Add constraint (7.6)
            else
                X̃_β ← Add constraint (7.5)
            end
        end
        x ← Solve the LP (7.1) with constraints X̃_β
        if infeasibility is found
            return no HC found
        elseif a HC is found
            return HC
        end
        splitting vertex ← Identify which vertex has 2 non-zero entries in x
        d ← Number of arcs emanating from splitting vertex
        for i from 1 to d
            G_d ← G with the d-th arc from splitting vertex fixed
            (G_d, new fixed arcs) ← Algorithm 7.1: Checking algorithm(G_d, fixed arcs)
            HC ← Algorithm 7.2: Branch and fix algorithm(G_d, β, new fixed arcs)
            if a HC is found
                return HC
            end
        end
        if a HC is found
            return HC
        else
            return no HC found
        end
end
```

Algorithm 7.2: Branch and fix algorithm

```
Input: G, β, fixed arcs
Output: function value

begin
        L(x) ← Sum of all arcs (1, j) emanating from vertex 1
        X̃_β ← constraints (4.18) and (4.20)
        for each arc in fixed arcs
            if arc goes into vertex 1
                X̃_β ← Add constraint (7.6)
            else
                X̃_β ← Add constraint (7.5)
            end
        end
        (x,function value) ← Solve the LP min L(x) subject to X_β
end
```

Algorithm 7.3: Second LP algorithm

Numerical results

We implement Algorithms 7.1–7.3 in MATLAB (version 7.4.0.336) and use CPLEX (version 11.0.0) to solve all the linear programming sub-problems. The algorithm is tested on a range of small- to medium-sized graphs. The results are encouraging. The number of branches required to solve each of these problems is only a tiny fraction of the number of deterministic policies. It is clear that non-Hamiltonian graphs require more branches to solve than Hamiltonian graphs of the same size. This is because in a Hamiltonian graph, as soon as a Hamiltonian cycle is found, the algorithm terminates. As there is no Hamiltonian cycle in a non-Hamiltonian graph, the algorithm only terminates after fathoming all generated branches of the logical B&F tree.

We provide a sample of results in Tables 7.1 and 7.2, including a comparison between the number of branches examined and the maximum possible number of branches (deterministic policies), and the running time in seconds. The Dodecahedron, Petersen, and Coxeter graphs, and the Knight's Tour problem are well-known in the literature (see Gross and Yellen [58, p. 12] for the first two, Gross and Yellen [58, p. 225] for the third, and Bondy and Murty [14, p. 241] for the last). The 24-vertex graph is a randomly chosen cubic 24-vertex graph.

Table 7.1: Preliminary results for the branch and fix method

Graph	**Branches**	**Upper bound**	**Time(s)**
Dodecahedron: Ham, $N = 20$, arcs $= 60$	75	3.4868×10^9	2.98
Ham, $N = 24$, arcs $= 72$	28	2.8243×10^{11}	1.02
8×8 Knight's Tour: Ham, $N = 64$, arcs $= 336$	failed	9.1654×10^{43}	> 12 hrs
Petersen: non-Ham, $N = 10$, arcs $= 30$	53	5.9049×10^4	0.99
Coxeter: non-Ham, $N = 28$, arcs $= 84$	11589	2.2877×10^{13}	593.40

In Table 7.2, in the first column we refer to the sets of cubic graphs with the prescribed number of vertices in the first column, in the second column we give the average number of branches examined by the branch and fix method, with the average taken over all graphs in the corresponding set of graphs. In the last three columns, we present the minimum and maximum branches examined over the set of graphs, and the average running time taken to solve the graphs in the corresponding class.

More specifically, we consider all 10-vertex cubic graphs, of which there are 17 Hamiltonian and 2 non-Hamiltonian graphs, and all 12-vertex cubic graphs, of which there are 80 Hamiltonian and 5 non-Hamiltonian graphs. We randomly generate 50 cubic graphs of size $N = 20, 30, 40$ and 50. All of the randomly

Table 7.2: Performance of the branch and fix method over cubic graphs

Type of graphs	Average branches	Minimum branches	Maximum branches	Average time(s)
Hamiltonian, N = 10	2.1	1	4	0.08
Hamiltonian, N = 12	3.4	1	10	0.14
Non-Hamiltonian, N = 10	32.5	12	53	0.61
Non-Hamiltonian, N = 12	25.6	11	80	0.53
50 graphs, N = 20	29.5	1	141	1.08
50 graphs, N = 30	216.5	3	1057	10.41
50 graphs, N = 40	2595.6	52	10536	160.09
50 graphs, N = 50	40316.7	324	232812	2171.46

generated graphs are Hamiltonian. See Meringer [79] for a reference on generating cubic graphs. We run every test with $\beta = 0.99$ and $\mu = 1/N^2$.

With this basic implementation of the B&F, we are not able to obtain a Hamiltonian solution for the 8×8 Knight's Tour problem after 12 hours. In Section 7.4, however, we introduce new constraints that allow us solve this problem in little more than a minute.

Example 7.1 *We describe a solution of the envelope graph (Figure 7.1), obtained using the aforementioned implementation of the B&F method, with* $\beta = 0.99$ *and* $\mu = 1/36$. *First, we solve the feasibility problem*

$$
\begin{aligned}
&x_{12} + x_{14} + x_{15} - \beta x_{21} - \beta x_{41} - \beta x_{51} = 1 - 5\mu,\\
&x_{21} + x_{23} + x_{26} - \beta x_{12} - \beta x_{32} - \beta x_{62} = \mu,\\
&x_{32} + x_{34} + x_{36} - \beta x_{23} - \beta x_{43} - \beta x_{63} = \mu,\\
&x_{41} + x_{43} + x_{45} - \beta x_{14} - \beta x_{34} - \beta x_{54} = \mu,\\
&x_{51} + x_{54} + x_{56} - \beta x_{15} - \beta x_{45} - \beta x_{65} = \mu,\\
&x_{62} + x_{63} + x_{65} - \beta x_{26} - \beta x_{36} - \beta x_{56} = \mu,\\
&\qquad\qquad x_{12} + x_{14} + x_{15} = \frac{(1-5\mu)(1-\beta) + \mu(\beta - \beta^6)}{(1-\beta)(1-\beta^6)},\\
&\qquad\qquad x_{ia} \geq 0, \quad \text{for all } (i,a) \in E(G).
\end{aligned}
$$

The first iteration produces a 1-randomised policy where the randomisation occurs at vertex 4. The logical B&F tree then splits into three choices: to fix arc (4,1), (4,3) or (4,5).

The algorithm first branches on fixing arc (4,1) (Figure 7.3). As the algorithm uses a depth-first search, arcs (4,3) and (4,5) will not be fixed unless the algorithm fathoms the (4,1) branch without having found a Hamiltonian cycle. Fixing the arc (4,1) is equivalent to eliminating arcs (4,3) and (4,5) in the remainder of this branch of the logical B&F tree. In addition, arcs (1,4),

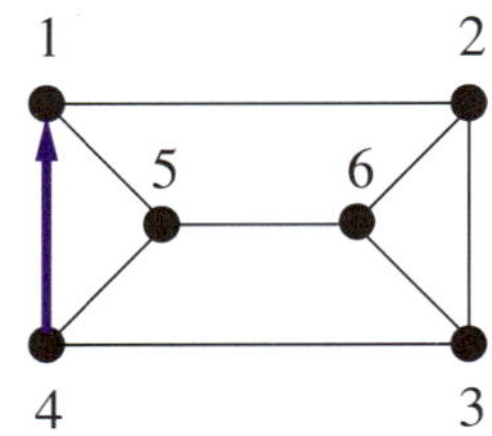

Fig. 7.3: Branching on arc $(4,1)$

(2,1) and (5,1) can also be eliminated because they cannot be present together with arc (4,1) in a Hamiltonian cycle.

At the second iteration we solve two LPs. We first solve the second LP, to check the feasibility of the graph remaining after the above round of fixing (and eliminating) of arcs

$$\min \{x_{12} + x_{15}\} \tag{7.11}$$

subject to

$$\begin{aligned} x_{12} + x_{15} - \beta x_{41} &= 1 - 5\mu, \\ x_{23} + x_{26} - \beta x_{12} - \beta x_{32} - \beta x_{62} &= \mu, \\ x_{32} + x_{34} + x_{36} - \beta x_{23} - \beta x_{63} &= \mu, \\ x_{41} - \beta x_{34} - \beta x_{54} &= \mu, \\ x_{54} + x_{56} - \beta x_{15} - \beta x_{65} &= \mu, \\ x_{62} + x_{63} + x_{65} - \beta x_{26} - \beta x_{36} - \beta x_{56} &= \mu, \\ -\beta^6 x_{12} - \beta^6 x_{15} + \beta x_{41} &= \frac{\mu(\beta - \beta^6)}{1-\beta}, \\ x_{ia} &\geq 0, \qquad \textit{for all } (i,a) \in E(G). \end{aligned}$$

The last equality constraint above comes from (7.6) because the fixed arc $(4,1)$ *returns to the home vertex. The optimal objective function returned is equal to the right-hand side of the omitted constraint (4.19), so we cannot fathom this branch, at this stage. Hence, we also solve the updated LP (7.1)*

$$\min \{-\beta^6 x_{12} - \beta^6 x_{15} + \beta x_{41}\}$$

subject to

$$\begin{aligned} x_{12} + x_{15} - \beta x_{41} &= 1 - 5\mu, \\ x_{23} + x_{26} - \beta x_{12} - \beta x_{32} - \beta x_{62} &= \mu, \\ x_{32} + x_{34} + x_{36} - \beta x_{23} - \beta x_{63} &= \mu, \end{aligned}$$

$$
\begin{aligned}
x_{41} - \beta x_{34} - \beta x_{54} &= \mu, \\
x_{54} + x_{56} - \beta x_{15} - \beta x_{65} &= \mu, \\
x_{62} + x_{63} + x_{65} - \beta x_{26} - \beta x_{36} - \beta x_{56} &= \mu, \\
x_{12} + x_{15} &= \frac{(1-5\mu)(1-\beta) + \mu(\beta - \beta^6)}{(1-\beta)(1-\beta^6)}, \\
x_{ia} &\geq 0, \quad \text{for all } (i,a) \in E(G).
\end{aligned}
$$

The second iteration produces a 1-randomised policy where the randomisation occurs at vertex 3. The logical B&F tree then splits into three choices: to fix arc (3,2), (3,4) or (3,6). The algorithm first selects the arc (3,2) to continue the branch. The graph at this stage is shown in Figure 7.4.

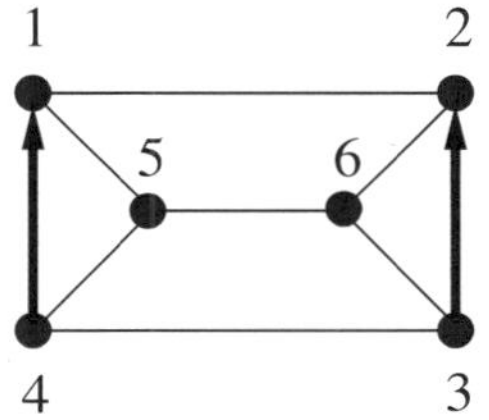

Fig. 7.4: Branching on arc $(3, 2)$ after fixing arc $(4, 1)$

Applying Checks 1–4 to the remaining vertices with multiple non-fixed arcs, we immediately see that arcs (2,6) and (1,5) must be fixed by Check 4. Once these arcs are fixed, arcs (5,4) and (6,3) are also fixed by Check 4. At this stage, every vertex has a fixed arc but we have not obtained a Hamiltonian cycle; hence, we fathom the branch. Travelling back up the tree, the algorithm next selects the arc (3,4) to branch on. Figure 7.5 shows the current graph.

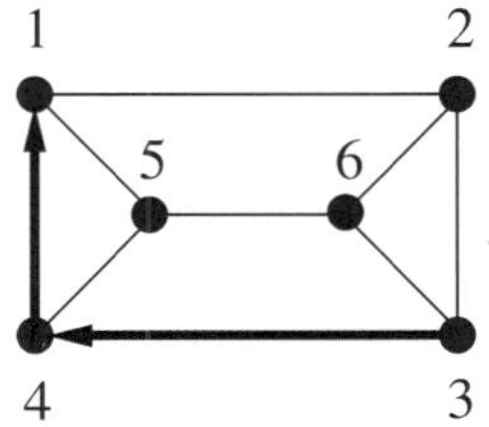

Fig. 7.5: Second branching on arc $(3, 4)$

Applying Checks 1–4 to the remaining vertices with multiple non-fixed arcs, we immediately see that arc (5,6) must be fixed by Check 3. Once this arc is fixed, arc (2,3) is also fixed by Check 3. Next, arc (6,2) is fixed by

Check 4 and, finally, arc (1,5) is fixed by Check 3. At this stage, every vertex has a fixed arc. Since these fixed arcs correspond to the Hamiltonian cycle $1 \to 5 \to 6 \to 2 \to 3 \to 4 \to 1$, *the algorithm terminates. The Hamiltonian cycle is shown in Figure 7.6.*

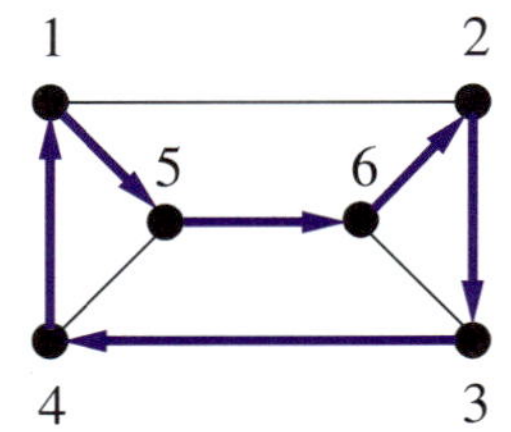

Fig. 7.6: Hamiltonian cycle found by the B&F method

The whole logical B&F tree is illustrated in Figure 7.7.

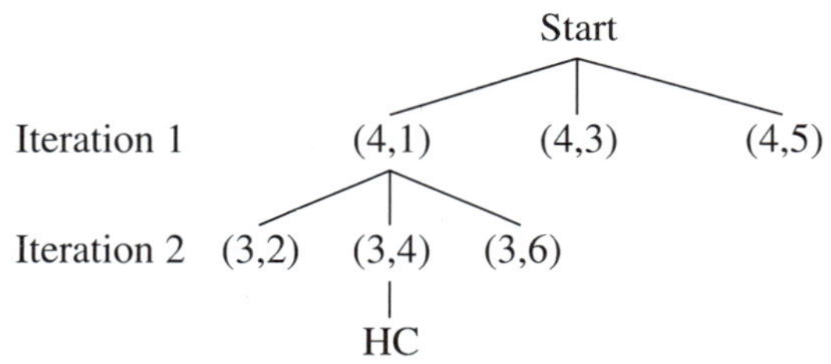

Fig. 7.7: The logical B&F tree for the envelope graph

We note that even in this simple example, in the worst case $3^6 = 729$ *branches are generated. However, our algorithm is able to find a Hamiltonian cycle after examining only two.*

Branching methods

A standard question when using a branching algorithm is which method of branching to use. A major benefit of the branch and fix method is that the checks and the second LP (7.9) often allow us to fathom a branching point relatively early, so depth-first searching is used. However, the horizontal ordering of the branch tree is, by default, determined by nothing more than the initial ordering of the vertices. For a non-Hamiltonian graph, this ordering makes no difference as the breadth of the entire tree will need to be traversed to determine that no Hamiltonian cycles exist. However, for a Hamiltonian graph, it is possible that a relabelling of the graph would result in the branch and fix method finding a Hamiltonian cycle sooner.

While it seems impossible to predict, in advance, which relabelling of vertices will find a Hamiltonian cycle the quickest, it is possible that the structure of the 1-randomised policy found at each branching point can provide information about which branch should be traversed first. Each 1-randomised policy with the splitting vertex i contains two non-zero values x_{ij} and x_{ik}, $j \neq k$. Without loss of generality, assume that $x_{ij} \leq x_{ik}$. We propose five branching methods:

1. Default branching (or vertex order): the branches are traversed in the order of the numerical labels of the vertices.

2. First branch on fixing (i, j), then (i, k) and then traverse the rest of the branches in vertex order.

3. First branch on fixing (i, k), then (i, j) and then traverse the rest of the branches in vertex order.

4. All branches are traversed in vertex order other than those corresponding to fixing (i, j) or (i, k). The last two branches traversed are (i, j) and then (i, k).

5. First branch on fixing (i, k), then traverse the rest of the branches other than the branch corresponding to fixing (i, j) in vertex order, and finally branch on fixing (i, j).

We test these five branching methods on the same sets of 50 randomly generated Hamiltonian cubic graphs as those generated for Table 7.2. In Table 7.3, we give the average number of branches examined by the B&F for each branching method.

Table 7.3: Average number of branches examined by the five branching methods over sets of Hamiltonian cubic graphs

Branching method	**20-vertex graphs**	**30-vertex graphs**	**40-vertex graphs**	**50-vertex graphs**
1	29.48	216.54	2595.58	40316.68
2	33.28	261.24	2227.24	43646.92
3	24.58	172.30	1624.50	17468.26
4	38.68	285.26	2834.44	53719.96
5	34.04	345.44	3228.44	76159.30

From the results shown in Table 7.3, it appears that Branching Method 3 is the best performing method for cubic graphs. The sets of cubic graphs are produced by GENREG [79], which uses a particular strategy of ordering the vertices that may also account for the success of Branching Method 3.

7.4 Wedge Constraints

Recall that constraints (4.18)–(4.20) that define $\widetilde{\mathcal{X}}_\beta$ depend upon parameters β and μ. While the use of β is necessary in the framework of a discounted Markov decision process, the selection of μ as a small, positive parameter is used only to ensure that the mapping $\boldsymbol{f}_x(i,a) = x_{ia}/x_i$ (see (4.17)) is well-defined. Without setting $\mu > 0$, it is possible for x_i to be 0. To illustrate this, recall constraint (4.18)

$$\sum_{i=1}^{N} \sum_{a \in \mathcal{A}(i)} \left(\delta_{ij} - \beta p(j|i,a)\right) x_{ia} = \nu_j, \quad \text{for } j \in \mathcal{S},$$

where

$$\nu_j = \begin{cases} 1-(N-1)\mu, & \text{if } i = 1, \\ \mu, & \text{otherwise.} \end{cases}$$

Rearranging constraint (4.18), we obtain

$$\sum_{a \in \mathcal{A}(j)} x_{ja} = \beta \sum_{i=1}^{N} \sum_{a \in \mathcal{A}(i)} p(j|i,a) x_{ia} + \nu_j, \quad \text{for } j \in \mathcal{S}.$$

Since we cannot ensure that

$$\sum_{i=1}^{N} \sum_{a \in \mathcal{A}(i)} p(j|i,a) x_{ia} \neq 0$$

for all j, we select $\mu > 0$ to ensure that $x_j = \sum_{a \in \mathcal{A}(j)} x_{ja} > 0$. However, an additional set of constraints, introduced in Eshragh *et al.* [43], can achieve the same goal while allowing us to set $\mu = 0$ by bounding x_j away from 0. We call these constraints *wedge constraints*. The wedge constraints are comprised of the following two sets of inequalities:

$$\sum_{a \in \mathcal{A}(i)} x_{ia} \leq \frac{\beta}{1-\beta^N}, \quad \text{for } i = 2, \ldots, N, \tag{7.12}$$

$$\sum_{a \in \mathcal{A}(i)} x_{ia} \geq \frac{\beta^{N-1}}{1-\beta^N}, \quad \text{for } i = 2, \ldots, N. \tag{7.13}$$

The rationale behind the wedge constraints is that when $\mu = 0$, we know from (7.2) that all Hamiltonian solutions to (4.18)–(4.20) take the form

$$x_{ia} = \begin{cases} \beta^k/(1-\beta^N) & \text{if } (i,a) \text{ is the } k\text{th arc on the HC,} \\ 0 & \text{otherwise.} \end{cases} \tag{7.14}$$

In every Hamiltonian cycle, exactly one arc emanates from each vertex. If (i_k, i_{k+1}) is an arc on a Hamiltonian cycle, then,

$$\sum_{a \in \mathcal{A}(i_k)} x_{i_k a} = x_{i_k i_{k+1}}. \tag{7.15}$$

Recall that we define the home vertex of a graph as vertex 1. Then, the initial (0th) arc in any Hamiltonian cycle is arc $(1, a)$, for some $a \in \mathcal{A}(1)$; therefore, from (7.14) we obtain

$$\sum_{a \in \mathcal{A}(1)} x_{1a} = \frac{1}{1-\beta^N}. \tag{7.16}$$

Constraint (7.16) is already given in (4.19) if we set $\mu = 0$. For all other vertices, however, constraints (7.12)–(7.13) will partially capture a new property of Hamiltonian solutions that is expressed in (7.14). In particular, substituting (7.14) into (7.15), for all vertices other than the home vertex, we obtain wedge constraints (7.12)–(7.13). Recall that in a cubic graph, there are exactly three arcs—say (i, a), (i, b) and (i, c)—from a given vertex i. Thus, in 3-dimensions, the corresponding constraints (7.12)–(7.13) have the shape indicated in Figure 7.8 that looks like a slice of a pyramid. The resulting wedge-like shape inspires the name wedge constraints.

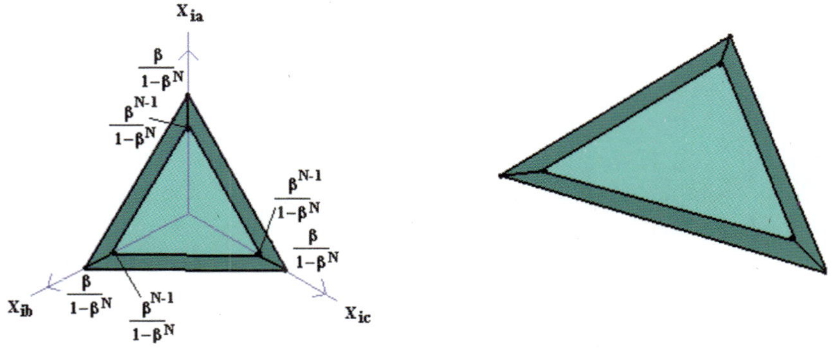

Fig. 7.8: Wedge constraints for a vertex in a cubic graph

We can add wedge constraints (7.12)–(7.13) to the constraint set (4.18)–(4.20), setting $\mu = 0$ in the latter. However, adding wedge constraints destroys the 1-randomised structure of non-Hamiltonian solutions that exists for the extreme points of the feasible region specified by (4.18)–(4.20), introducing many new extreme points to the feasible region. Since this is undesirable, we only use wedge constraints when solving the second LP (with the objective function (7.9)), in an attempt to determine whether a branch can be fathomed earlier than is the case without the wedge constraints.

The model incorporating the wedge constraints is

$$\sum_{i=1}^{N} \sum_{a \in \mathcal{A}(i)} \left(\delta_{ij} - \beta p(j|i,a)\right) x_{ia} = \delta_{1j}, \quad \text{for } j \in \mathcal{S}, \tag{7.17}$$

$$x_{ia} \geq 0, \quad \text{for } (i,a) \in E(G), \tag{7.18}$$

$$\sum_{a \in \mathcal{A}(i)} x_{ia} \leq \frac{\beta}{1-\beta^N}, \quad \text{for } i = 2, \ldots, N, \tag{7.19}$$

$$\sum_{a \in \mathcal{A}(i)} x_{ia} \geq \frac{\beta^{N-1}}{1-\beta^N}, \quad \text{for } i = 2, \ldots, N, \tag{7.20}$$

which replaces constraints (4.18) and (4.20) in the second LP (7.9). Everything else in the branch and fix method is identical to the method described in Section 7.2, except that the obtained value of the objective function (7.9) is now compared to the right-hand side of (7.16), that is, the right-hand side of (4.19) with μ set to 0.

We run this model on the same selection of graphs as those in Tables 7.1–7.3, to compare its performance to that of the original branch and fix method. There is a significant decrease in the number of branches examined, and consequently in the running time of the model. Tables 7.4–7.6 show a sample of results. We run all tests with $\beta = 0.99$, and $\mu = 1/N^2$. In Table 7.4, we compare the number of branches examined by the B&F for five graphs to the maximum possible number of branches (deterministic policies), and show the running time in seconds. In Table 7.5, we solve several sets of cubic graphs, and report the average number of branches examined by the B&F over each set. We also report the minimum and maximum branches examined over each set, and the average running time. As in Table 7.2, we consider all 10-vertex cubic graphs, of which there are 17 Hamiltonian and 2 non-Hamiltonian graphs, and all 12-vertex cubic graphs, of which there are 80 Hamiltonian and 5 non-Hamiltonian graphs. For each set of larger cubic graphs, we use the same graphs as were randomly generated for Table 7.2. Table 7.6 shows the average number of branches examined by the B&F for the same set of randomly cubic graphs shown in Table 7.5, for all five branching methods.

Table 7.4: Preliminary results for the branch and fix method with wedge constraints included

Graph	**Branches**	**Upper bound**	**Time(s)**
Dodecahedron: Ham, $N = 20$, arcs $= 60$	43	3.4868×10^9	1.71
Ham, $N = 24$, arcs $= 72$	5	2.8243×10^{11}	0.39
8×8 Knight's Tour: Ham, $N = 64$, arcs $= 336$	220	9.1654×10^{43}	78.46
Petersen: non-Ham, $N = 10$, arcs $= 30$	53	5.9049×10^4	1.17
Coxeter: non-Ham, $N = 28$, arcs $= 84$	5126	2.2877×10^{13}	262.28

Table 7.5: Performance of the branch and fix method with wedge constraints included over cubic graphs

Type of graphs	**Average branches**	**Minimum branches**	**Maximum branches**	**Average time(s)**
Hamiltonian, N = 10	2.1	1	4	0.09
Hamiltonian, N = 12	3.0	1	10	0.14
Non-Hamiltonian, N = 10	32.5	12	53	0.70
Non-Hamiltonian, N = 12	23.2	11	72	0.50
50 graphs, N = 20	14.6	1	75	0.65
50 graphs, N = 30	41.7	2	182	2.56
50 graphs, N = 40	209.2	7	1264	18.11
50 graphs, N = 50	584.4	8	2522	67.99

Table 7.6: Average number of branches examined by the five branching methods with wedge constraints included over sets of Hamiltonian cubic graphs

Branching method	**20-vertex graphs**	**30-vertex graphs**	**40-vertex graphs**	**50-vertex graphs**
1	14.60	41.66	209.18	584.40
2	14.76	49.94	164.74	636.88
3	11.86	38	145.54	603.24
4	17.28	50.32	152.50	632.08
5	15.32	39.42	123.72	356.32

We observe that for smaller-size graphs, wedge constraints do not improve the performance of the B&F method. For larger-size graphs, however, a significant improvement is evident. This is especially striking for graphs with 30–40 vertices. We also note with interest that the model with wedge constraints included performs well when we select Branching Method 5, which is the worst performing branching method in the model without wedge constraints. Since Branching Method 3 also performs well, and both Branching Methods 3 and 5 involve branching first on arc (i, k), it appears that branching first on arc (i, k) is an efficient strategy for graphs generated by GENREG when we include wedge constraints.

In Andramonov *et al.* [4], the first numerical procedure taking advantage of the MDP embedding is used to solve graphs of similar sizes to those listed in Tables 7.4 and 7.5. The model given in [4] is solved using the Mix Integer Programming (MIP) solver in CPLEX 4.0. The present model outperforms the model in [4] in terms of the number of branches examined in the cases displayed in Tables 7.4 and 7.5. In particular, the number of branches examined to solve each graph is much reduced in the branch and fix method when compared to similar sized graphs in [4]. In particular, to solve the 8×8 Knight's Tour graph, between 2000 and 40000 branches were examined in the model given in [4] (depending on the selection of parameters in CPLEX), but only 220 were required in the branch and fix method with wedge constraints included. This significant improvement highlights both the progress made in this line of research over the last decade, and the advantages obtained by the use of wedge constraints. We seek to take further advantage of the wedge constraints in a new mixed integer programming formulation, *the Wedged-MIP heuristic*, which we describe in the next section.

7.5 The Wedged-MIP heuristic

The discussion in the preceding section naturally leads us to consider the polytope $\mathcal{Y}_\beta$ defined by the following seven sets of linear constraints

$$\sum_{i=1}^{N} \sum_{a \in \mathcal{A}(i)} (\delta_{ij} - \beta p(j|i,a))\, y_{ia} = \delta_{1j}(1-\beta^N), \quad \text{for } j \in \mathcal{S}, \tag{7.21}$$

$$\sum_{a \in \mathcal{A}(1)} y_{1a} = 1, \tag{7.22}$$

$$y_{ia} \geq 0, \quad \text{for } (i,a) \in E(G), \tag{7.23}$$

$$\sum_{j \in \mathcal{A}(i)} y_{ij} \leq \beta, \quad \text{for } i \in \mathcal{S}\backslash\{1\}, \tag{7.24}$$

$$\sum_{j \in \mathcal{A}(i)} y_{ij} \geq \beta^{N-1}, \quad \text{for } i \in \mathcal{S}\backslash\{1\}, \tag{7.25}$$

$$y_{ij} + y_{ji} \leq 1, \quad \text{for } (i,j),(j,k) \in E(G), \tag{7.26}$$

$$y_{ij} + y_{jk} + y_{ki} \leq 2, \quad \text{for } (i,j),(j,k),(k,i) \in E(G). \tag{7.27}$$

Remark 7.1.

1. The variables y_{ia} that define $\mathcal{Y}_\beta$ are obtained from the variables x_{ia} that define $\mathcal{X}_\beta$ by the transformation

$$y_{ia} = (1-\beta^N)x_{ia}, \quad \text{for } (i,a) \in E(G).$$

2. In view of Item 1, constraints (7.21)–(7.23) are merely constraints (4.18)–(4.20) normalised by the multiplier $(1-\beta^N)$, and with $\mu = 0$. Furthermore, constraints (7.25)–(7.24) are similarly normalised wedge constraints (7.12)–(7.13).

3. Note that normalising (7.14) in the same way implies that if $\boldsymbol{y} \in \mathcal{Y}_\beta$ corresponds to a Hamiltonian cycle, then
$$y_{ia} \in \{0, 1, \beta, \ldots, \beta^{N-1}\}, \qquad \text{for } (i,a) \in E(G).$$
Thus, for β sufficiently near 1 all positive entries of a Hamiltonian solution $\boldsymbol{y}$ are also either 1 (if $i = 1$), or close to 1. Therefore, if $\boldsymbol{y} \in \mathcal{Y}_\beta$ is any feasible point with only one positive entry y_{ia} for all $a \in \mathcal{A}(i)$, for each i, then constraints (7.25)–(7.24) ensure that all those positive entries have values near 1. Furthermore, constraints (7.26)–(7.27) ensure that at most one such large entry is permitted on any potential 2-cycle, and at most two such large entries are permitted on any potential 3-cycle.

4. In view of Item 3, it is reasonable to search for a feasible point $\boldsymbol{y} \in \mathcal{Y}_\beta$ that has only a single positive entry y_{ia} for all $a \in \mathcal{A}(i)$ and for each i.

We make the last point of the above remark precise in the following proposition that is analogous to a result proved in Chen and Filar [22] for an embedding of the HCP in a long-run average MDP. Since it forms the theoretical basis of our most powerful heuristic, for the sake of completeness, we supply a formal proof below.

Proposition 7.1. *Given any graph G and its embedding in a discounted Markov decision process, consider the polytope $\mathcal{Y}_\beta$ defined by (7.21)–(7.27) for $\beta \in [0,1)$ and sufficiently near 1. The following statements are equivalent:*

(i) The point $\hat{\boldsymbol{y}} \in \mathcal{Y}_\beta$ is Hamiltonian in the sense that the positive entries $\hat{y}_{ia}$ of $\hat{\boldsymbol{y}}$ correspond to arcs (i,a) defining a Hamiltonian cycle in G.

(ii) The point $\hat{\boldsymbol{y}} \in \mathcal{Y}_\beta$ is a global minimiser of the nonlinear program
$$\begin{aligned} &\min \sum_{i=1}^{N} \sum_{a \in \mathcal{A}(i)} \sum_{b \in \mathcal{A}(i), b \neq a} y_{ia} y_{ib} \\ &\textit{subject to} \\ &\boldsymbol{y} \in \mathcal{Y}_\beta. \end{aligned} \tag{7.28}$$
which gives the objective function value of 0 in (7.28).

(iii) The point $\hat{\boldsymbol{y}} \in \mathcal{Y}_\beta$ satisfies the additional set of nonlinear constraints
$$y_{ia} y_{ib} = 0, \quad \textit{for } i \in \mathcal{S}, a, b \in \mathcal{A}(i), a \neq b, \tag{7.29}$$

Proof. By the nonnegativity of $\hat{\boldsymbol{y}} \in \mathcal{Y}_\beta$, it immediately follows that Parts (ii) and (iii) are equivalent.

From (7.14) and Item 3 of Remark 7.1 we note that if $\hat{\boldsymbol{y}}$ is Hamiltonian, it implies Part (iii). Furthermore, if $\hat{\boldsymbol{y}}$ is a global minimiser of (7.28), constraints (7.25) ensure that it must have at least one positive entry corresponding to some arc $a \in \mathcal{A}(i)$ for each $i \in \mathcal{S}$. Since $\hat{y}_{ia}\hat{y}_{ib} = 0$ for all $a \neq b$, $i \in \mathcal{S}$, we conclude that $\hat{\boldsymbol{y}}$ has exactly one positive entry $\hat{y}_{ia}$, for each i. We then define $\hat{\boldsymbol{x}} \in \widetilde{\mathcal{X}}_\beta$ by $\hat{x}_{ia} = 1/(1-\beta^N)\hat{y}_{ia}$ for all $(i,a) \in E(G)$, and use (7.14) to construct the policy $\hat{\boldsymbol{f}} = M^{-1}(\hat{\boldsymbol{x}})$. It is clear that $\hat{\boldsymbol{f}} \in \mathcal{F}_\mathcal{D}$, and hence $\hat{\boldsymbol{x}}$ is Hamiltonian by Part (ii) of Proposition 4.2. Since positive entries of $\hat{\boldsymbol{x}}$ and $\hat{\boldsymbol{y}}$ occur at precisely the same arcs (i,a), $\hat{\boldsymbol{y}}$ is also Hamiltonian, so Part (iii) implies Part (i). □

Corollary 7.2. *If $\mathcal{Y}_\beta = \emptyset$, the empty set, then the graph G is non-Hamiltonian. If $\mathcal{Y}_\beta \neq \emptyset$, the (possibly empty) set of Hamiltonian solutions $\mathcal{Y}_\beta^{\mathrm{H}} \subset \mathcal{Y}_\beta$ is in one-to-one correspondence with Hamiltonian cycles of G, and satisfies*

$$\mathcal{Y}_\beta^{\mathrm{H}} = \mathcal{Y}_\beta \cap \{\boldsymbol{y} \mid (7.29) \textit{ holds}\}.$$

Proof. By construction, if G is Hamiltonian, then there exists a policy $\hat{\boldsymbol{f}} \in \mathcal{F}_\mathcal{D}$ tracing out a Hamiltonian cycle in G. Let $\hat{\boldsymbol{x}} = M(\hat{\boldsymbol{f}})$ using (4.13) with $\boldsymbol{\nu} = \boldsymbol{e}_1^{\mathrm{T}}$, and define $\hat{\boldsymbol{y}} = (1-\beta^N)\hat{\boldsymbol{x}}$. Clearly, $\hat{\boldsymbol{y}} \in \mathcal{Y}_\beta$, so $\mathcal{Y}_\beta \neq \emptyset$. From Proposition 7.1, it follows that $\hat{\boldsymbol{y}}$ satisfies (7.29). Conversely, only points in $\mathcal{Y}_\beta^{\mathrm{H}}$ define Hamiltonian solutions in $\mathcal{Y}_\beta$. □

For symmetric graphs, in which arc $(i,a) \in E(G)$ if and only if arc $(a,i) \in E(G)$, we further improve the wedge constraints (7.25)–(7.24) by considering the shortest path between the home vertex and each other vertex in the graph. We define $\ell(i,j)$ to be the length of the shortest path between vertices i and j in G.

Lemma 7.1. *Any Hamiltonian solution to (7.21)–(7.23) satisfies the following constraints*

$$\sum_{a\in\mathcal{A}(i)} y_{ia} \leq \beta^{\ell(1,i)}, \quad \textit{for } i \in \mathcal{S}\backslash\{1\}, \tag{7.30}$$

$$\sum_{a\in\mathcal{A}(i)} y_{ia} \geq \beta^{N-\ell(1,i)}, \quad \textit{for } i \in \mathcal{S}\backslash\{1\}. \tag{7.31}$$

Proof. From (7.14) and Item 3 of Remark 7.1, we know that for a Hamiltonian cycle in which the kth arc is the arc (i,a), the corresponding variable $y_{ia} = \beta^k$, and all other $y_{ib} = 0$, $b \neq a$. Therefore,

$$\sum_{a\in\mathcal{A}(i)} y_{ia} = \beta^k. \tag{7.32}$$

Then, since it takes at least $\ell(1,i)$ arcs to reach vertex i from the home vertex 1, we immediately obtain that $k \geq \ell(1,i)$, and therefore

$$\sum_{a \in \mathcal{A}(i)} y_{ia} \leq \beta^{\ell(1,i)},$$

which coincides with (7.30). Since G is an undirected graph, we know that $\ell(i,1) = \ell(1,i)$. Thus, we obtain that $k \leq N - \ell(1,i)$, and hence,

$$\sum_{a \in \mathcal{A}(i)} y_{ia} \geq \beta^{N-\ell(1,i)},$$

which coincides with (7.31). □

Given the above, we reformulate the HCP as a mixed (non-linear) integer programming feasibility problem, which we call the *Wedged-MIP heuristic*, as follows

$$\sum_{i=1}^{N} \sum_{a \in \mathcal{A}(i)} (\delta_{ij} - \beta p(j|i,a))\, y_{ia} = \delta_{1j}(1-\beta^N), \quad \text{for } j \in \mathcal{S}, \tag{7.33}$$

$$\sum_{a \in \mathcal{A}(1)} y_{1a} = 1, \tag{7.34}$$

$$y_{ia} \geq 0, \quad \text{for } (i,a) \in E(G), \tag{7.35}$$

$$\sum_{j \in \mathcal{A}(i)} y_{ij} \leq \beta^{\ell(1,i)}, \quad \text{for } i \in \mathcal{S}\backslash\{1\}, \tag{7.36}$$

$$\sum_{j \in \mathcal{A}(i)} y_{ij} \geq \beta^{N-\ell(1,i)}, \quad \text{for } i \in \mathcal{S}\backslash\{1\}, \tag{7.37}$$

$$y_{ij} + y_{ji} \leq 1, \quad \text{for } (i,j),(j,i) \in E(G), \tag{7.38}$$

$$y_{ij} + y_{jk} + y_{ki} \leq 2, \quad \text{for } (i,j),(j,k),(k,i) \in E(G), \tag{7.39}$$

$$y_{ia} y_{ib} = 0, \quad \text{for } i \in \mathcal{S}, a,b \in \mathcal{A}(i), a \neq b. \tag{7.40}$$

We solve the above formulation in IBM ILOG OPL-CPLEX 5.1. A benefit of this solver is that the non-linear constraints (7.40) may be submitted in a format usually not acceptable in CPLEX and the IBM ILOG OPL-CPLEX CP Optimizer will interpret them in a way suitable for CPLEX. We allow these constraints to be submitted in one of two different ways, left up to the user's choice. We define the operator == as follows

$$(a == b) = \begin{cases} 1 & \text{if } a = b, \\ 0 & \text{otherwise,} \end{cases}$$

the operator $!=$ as follows

$$(a \,!= b) = \begin{cases} 0 & \text{if } a = b, \\ 1 & \text{otherwise,} \end{cases}$$

and denote by d_i the number of arcs emanating from vertex i. Then, we submit constraints (7.40) to IBM ILOG OPL-CPLEX in either of the forms

$$\sum_{a \in \mathcal{A}(i)} (y_{ia} == 0) = d_i - 1, \quad \text{for } i \in \mathcal{S}, \tag{7.41}$$

or

$$\sum_{a \in \mathcal{A}(i)} (y_{ia} \,!= 0) = 1, \quad \text{for } i \in \mathcal{S}. \tag{7.42}$$

Even though constraints (7.41) and (7.42) are theoretically identical when added to (7.33)–(7.39), their interpretation by IBM ILOG OPL-CPLEX produces different solutions, with different running times. Neither choice solves graphs consistently faster than the other, so if one form fails to find a solution quickly, we try the other form. Using this model we are able to obtain Hamiltonian solutions efficiently for many large graphs, using a Pentium 3.4GHz with 4GB RAM. Table 7.7 presents results for eight graphs.

Table 7.7: Running times for Wedged-MIP heuristic

Graph	N	**Arcs**	β	**Time(hh:mm:ss)**
8×8 Knight's Tour	64	336	0.99999	00:00:02
Perturbed Horton	94	282	0.99999	00:00:02
12×12 Knight's Tour	144	880	0.99999	00:00:03
250-vertex	250	1128	0.99999	00:00:16
20×20 Knight's Tour	400	2736	0.99999	00:20:57
500-vertex	500	3046	0.99999	00:10:01
1000-vertex	1000	3996	0.999999	00:30:46
2000-vertex	2000	7992	0.999999	10:24:05

We note that the perturbed Horton graph given here is a 94-vertex cubic graph that, unlike the original Horton graph (96-vertex cubic graph [103]), is Hamiltonian. The 250- and 500-vertex graphs are both non-regular graphs that are randomly generated for testing purposes, while the 1000- and 2000-vertex graphs come from the TSPLIB website, maintained by University of Heidelberg [98].

We present a visual representation of a solution to the 250-vertex graph found by the Wedged-MIP heuristic in Figure 7.9, where the vertices are drawn as blue dots clockwise in an ellipse, with vertex 1 at the top, and the arcs between the vertices are inside the ellipse. The arcs in the Hamiltonian cycle found by the Wedged-MIP heuristic are highlighted red, and all other arcs are shown in blue. While it is very difficult to make out much detail from Figure 7.9, it serves as a good illustration of the complexity involved in problems of this size.

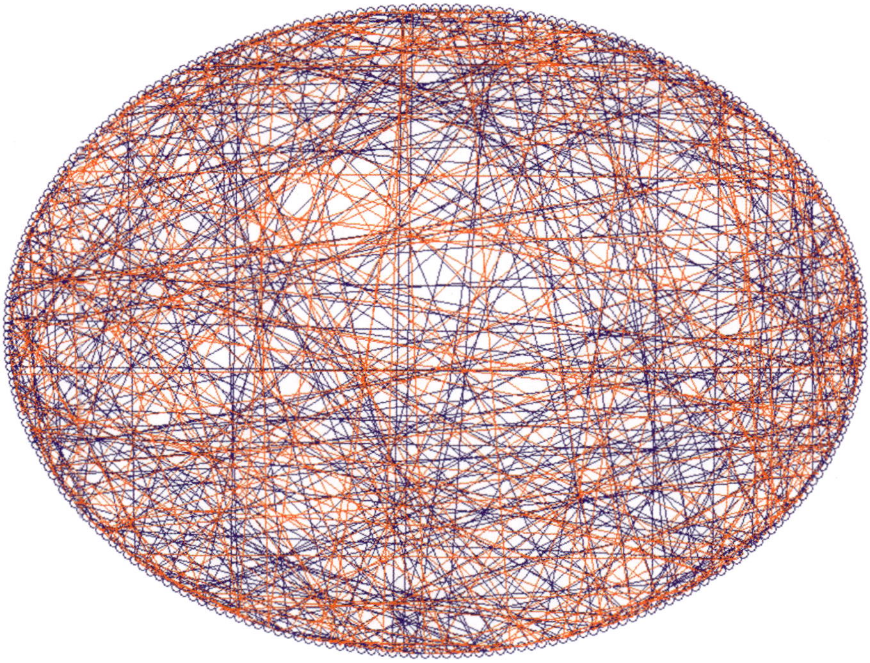

Fig. 7.9: Solution to 250-vertex graph (Hamiltonian cycle in red)

Comparisons between Wedged-MIP heuristic and two TSP formulations

Since we solve the Wedged-MIP heuristic in OPL-CPLEX, we investigate two other, well-known, MIP formulations, and also solve them in OPL-CPLEX for the same set of graphs as those in Table 7.7, as well as three randomly generated cubic Hamiltonian graphs of sizes 12, 24 and 38, as a benchmark test. The two other formulations are the *modified single commodity flow model* [54] and the *third stage dependent model* [101]. We select these two formulations because they are the best performing methods of each type (commodity flow and stage dependent, respectively) reported in Orman and Williams [83]. Both the modified single commodity flow and the third stage dependent models were designed to solve the travelling salesman problem, so they contain costs/distances c_{ij} for each arc $(i,j) \in E(G)$. Since we only want to solve the HCP with these formulations, we set

$$c_{ij} = \begin{cases} 1 & \text{for } (i,j) \in E(G), \\ 0 & \text{for } (i,j) \notin E(G). \end{cases}$$

The modified single commodity flow model (MSCF) is formulated with decision variables x_{ij} and y_{ij} as follows

$$\min \sum_{(i,j)\in E(G)} c_{ij}x_{ij}$$

subject to

$$\begin{aligned}
\sum_{j\in\mathcal{A}(i)} x_{ij} &= 1, \quad \text{for } i \in \mathcal{S}, \\
\sum_{i\in\mathcal{B}(j)} x_{ij} &= 1, \quad \text{for } j \in \mathcal{S}, \\
\sum_{j\in\mathcal{A}(1)} y_{1j} &= N-1, \\
\sum_{i\in\mathcal{B}(j)} y_{ij} - \sum_{k\in\mathcal{A}(j)} y_{jk} &= 1, \quad \text{for } j \in \mathcal{S}\backslash\{1\},
\end{aligned}$$

$$\begin{aligned}
y_{1j} &\le (N-1)x_{1j}, \quad \text{for } j \in \mathcal{A}(1), \\
y_{ij} &\le (N-2)x_{ij}, \quad \text{for } i \in \mathcal{S}\backslash\{1\}, j \in \mathcal{A}(i), \\
x_{ij} &\in \{0,1\}, \\
y_{ij} &\ge 0.
\end{aligned}$$

The third stage dependent model (TSD) is formulated with decision variables x_{ij} and $y_{ij}^t, t \in \mathcal{S}$, as follows

$$\min \sum_{(i,j)\in E(G)} c_{ij}x_{ij}$$

subject to

$$\begin{aligned}
\sum_{j\in\mathcal{A}(i)} x_{ij} &= 1, \quad \text{for } i \in \mathcal{S}, \\
\sum_{i\in\mathcal{B}(j)} x_{ij} &= 1, \quad \text{for } j \in \mathcal{S},
\end{aligned}$$

$$
\begin{aligned}
x_{ij} - \sum_{t=1}^{N} y_{ij}^t &= 0, \quad \text{for } (i,j) \in E(G),\\
\sum_{t=1}^{N} \sum_{j \in \mathcal{A}(i)} y_{ij}^t &= 1, \quad \text{for } i \in \mathcal{S},\\
\sum_{t=1}^{N} \sum_{i \in \mathcal{B}(j)} y_{ij}^t &= 1, \quad \text{for } j \in \mathcal{S},\\
\sum_{(i,j) \in E(G)} y_{ij}^t &= 1, \quad \text{for } t \in \mathcal{S},\\
\sum_{j \in \mathcal{A}(1)} y_{1j}^1 &= 1,\\
\sum_{i \in \mathcal{B}(1)} y_{i1}^N &= 1,\\
\sum_{j \in \mathcal{A}(i)} y_{ij}^t - \sum_{k \in \mathcal{B}(i)} y_{ki}^{t-1} &= 0, \quad \text{for } i \in \mathcal{S}, \text{ and } t \in \mathcal{S} \backslash \{1\}.
\end{aligned}
$$

We run these two models on the same graphs as shown in Table 7.7, as well as three additional randomly generated cubic Hamiltonian graphs of orders 12, 24 and 38. Table 7.8 shows the running times of these two models, along with the Wedged-MIP heuristic running times. For each graph tested other than the 1000- and 2000-vertex graph, we run MCSF and TSD for 24 hours, and if no solution is found we terminate the execution. For the 1000- and 2000-vertex graphs, we allow 168 hours (1 week) before terminating.

Table 7.8: Running times (hh:mm:ss) for Wedged-MIP heuristic, MCSF and TSD

Graph	N	**Wedged-MIP heuristic**	**MCSF**	**TSD**
12-vertex cubic	12	00:00:01	00:00:01	00:00:01
24-vertex cubic	24	00:00:01	00:00:01	00:00:02
38-vertex cubic	38	00:00:01	00:00:01	00:21:04
8×8 Knight's Tour	64	00:00:02	00:00:01	$>$ 24 hours
Perturbed Horton	94	00:00:02	00:03:04	$>$ 24 hours
12×12 Knight's Tour	144	00:00:03	00:01:12	$>$ 24 hours
250-vertex	250	00:00:16	00:29:42	$>$ 24 hours
20×20 Knight's Tour	400	00:20:57	17:35:57	$>$ 24 hours
500-vertex	500	00:10:01	$>$ 24 hours	$>$ 24 hours
1000-vertex	1000	00:30:46	$>$ 1 week	$>$ 1 week
2000-vertex	2000	10:24:05	$>$ 1 week	$>$ 1 week

In summary, in this chapter, we considered the space $\mathcal{X}_\beta$ of discounted occupational measures associated with the embedding of a given graph in a discounted MDP as described in Chapter 4. We demonstrated that when modified by a small number of additional constraints, the resulting space forms a suitable domain for a number of promising algorithmic approaches to the HCP. We feel that this way of tackling the HCP, numerically, can be explored much further. In particular, for graphs with large number of vertices, using a value of β that is very close to 1 could present accuracy problems in verifying validity of linear constraints involving that parameter. It is possible that exploiting techniques similar to those used to develop the parameter-free model of Section 4.5 may help resolve this problem.

Chapter 8
Interior Point and Cross-Entropy Algorithms

8.1 Introduction

In this chapter, we briefly discuss two recent algorithms that exploit two modern trends in optimisation in the context of our stochastic embedding of the Hamiltonian cycle problem: the *interior point method* and the *importance sampling method*. In particular, the first algorithm searches in the interior of the convex domain of doubly stochastic matrices induced by a given graph, with the goal of converging to an extreme point corresponding to a permutation matrix that coincides with a Hamiltonian cycle. The search is carried out with the help of a suitably-adapted *barrier method* (den Hertog [63]). The second algorithm is a hybrid algorithm that assigns nominal costs to edges in a given graph, and augments the graph with artificial edges (with high costs) to form a complete graph as the base of an auxiliary instance of the travelling salesman problem. Since it is easy to generate Hamiltonian cycles in the resulting graph, the *cross-entropy* technique of Rubinstein and Kroese [95] cleverly exploits importance sampling-type techniques, in order to generate elite sample points that correspond to fully-randomised probability transition matrices that begin favouring edges with low costs—the original, *authentic edges* in the given graph. However, instead of waiting for this process to converge, the presented algorithm switches to a linear programming problem on a suitable subspace of discounted occupational measures where the coefficients of the objective function are the entries of the best probability transition matrix generated so far.

In Chapter 5, we characterised Hamiltonian cycles as global maxima of suitably-constructed determinants over the space of probability transition matrices of stationary policies in our Markov decision process. These results can be used as a basis for alternative algorithmic approaches. In this section, we outline one possible approach that exploits the LU-decomposition of $\mathbf{I}-\mathbf{P}$, where $\mathbf{P}$ is the probability transition matrix in the interior of the set of doubly stochastic matrices induced by any given graph. These decompositions greatly simplify the computational effort that would otherwise be needed in implementations of interior point (logarithmic barrier)-type heuristics for finding a Hamiltonian cycle. We discuss below the main steps of such a heuristic and its numerical performance. See the PhD thesis of Haythorpe [62] for more detail.

8.2 Interior Point Method Algorithm

Consider a constrained optimisation problem of the form

$$\begin{aligned} &\min f(\boldsymbol{x}) \\ \text{subject to}& \\ &h_i(\boldsymbol{x}) \geq 0, \qquad \text{for } i = 1, \ldots, m, \end{aligned} \tag{8.1}$$

that has a solution $\boldsymbol{x}^* \in \mathbb{R}^n$. We assume that the feasible region Ω has a nonempty interior, denoted by $\mathrm{Int}(\Omega)$; that is, there exists an $\boldsymbol{x}$ such that $h_i(\boldsymbol{x}) > 0$ for $i = 1, 2, \ldots, m$. We also assume that $f(\boldsymbol{x})$ and $h_i(\boldsymbol{x})$, for $i = 1, 2, \ldots, m$, are continuous functions and possess derivatives up to order 2. This problem is often solved by an interior point method. One such method is to consider a parameterized, auxiliary objective function of the form

$$F(\boldsymbol{x}) = f(\boldsymbol{x}) - \mu \sum_{i=1}^{m} \ln(h_i(\boldsymbol{x})), \tag{8.2}$$

for some $\mu > 0$, and an associated, essentially unconstrained, auxiliary optimisation problem

$$\begin{aligned} &\min F(\boldsymbol{x}) \\ \text{subject to}& \\ &\boldsymbol{x} \in \mathrm{Int}(\Omega). \end{aligned} \tag{8.3}$$

For each inequality constraint, the auxiliary objective function $F(\boldsymbol{x})$ contains a logarithmic term, which ensures that $F(\boldsymbol{x})$ increases to infinity as $h_i(\boldsymbol{x})$ decreases to zero. The sum of logarithmic terms ensures that any minimiser of $F(\boldsymbol{x})$ strictly satisfies the inequality constraints in the optimisation model (8.1) for $\mu > 0$. If we choose μ large enough that $F(x)$ is strictly convex, then its unique global minimiser $\boldsymbol{x}^*(\mu)$ is well-defined. In well-behaved cases, it is usually expected that as μ decreases to zero, the limit of $\boldsymbol{x}^*(\mu)$ exists and constitutes a global minimum to (8.1).

We define a sequence $\{\mu_k\}_{k=0}^{\infty}$ such that $\mu_k > 0$ for all k and the sequence converges to zero. We associate with this sequence a set of auxiliary objective functions

$$F_k(\boldsymbol{x}) = f(\boldsymbol{x}) - \mu_k \sum_{i=1}^{m} \ln(h_i(\boldsymbol{x})),$$

which has a sequence of minimisers $\{\boldsymbol{x}^k\}$. It is well-known that if $f(\boldsymbol{x})$ is convex and $h_i(\boldsymbol{x})$ are concave for all $i = 1, 2, \ldots, m$, then $\boldsymbol{x}^k$ converges to $\boldsymbol{x}^*$ (see den Hertog [63, pp. 49–65]). In our implementation, we select $\mu_{k+1} = 0.9\mu_k$, which is an arbitrary choice but generally performs well. A different multiplier besides 0.9 may be used, or left as an input parameter. We note that although in our application $f(\boldsymbol{x})$ is non-convex, it is still reasonable to expect an interior point method such as the above to perform well as a heuristic.

We can add linear equality constraints to the optimisation model (8.1) to form a new problem:

$$\begin{aligned} &\min f(\boldsymbol{x}) \\ \text{subject to} & \\ &h_i(\boldsymbol{x}) \geq 0, \quad \text{for } i = 1, \ldots, m, \\ &\mathbf{Q}\boldsymbol{x} = \boldsymbol{b}. \end{aligned} \tag{8.4}$$

Once we obtain an initial interior point satisfying the constraints in (8.4), it is possible to convert the minimisation problem (8.4) to the minimisation problem (8.1) by working in the null space of the equality constraints $\mathbf{Q}\boldsymbol{x} = \boldsymbol{b}$.

We define the auxiliary objective function $F(\boldsymbol{x})$ as in (8.2), and denote by the matrix $\mathbf{Z}$ the null space of the linear equality constraints, so $\mathbf{QZ} = \mathbf{0}$ and the columns of $\mathbf{Z}$ are linearly independent. Then, given the current feasible point $\boldsymbol{x}^k$, we find a search direction $\boldsymbol{d}^k$ in the null space $\mathbf{Z}$, by solving for $\boldsymbol{d}^k$ and $\overline{\boldsymbol{y}}$ the system

$$\mathbf{H}(\boldsymbol{x}^k)\boldsymbol{d}^k = \mathbf{Q}^{\mathrm{T}}\overline{\boldsymbol{y}} - \boldsymbol{g}(\boldsymbol{x}^k), \tag{8.5}$$

$$\mathbf{Q}\boldsymbol{d}^k = \mathbf{0}, \tag{8.6}$$

where $\boldsymbol{g}(\boldsymbol{x}^k)$ and $\mathbf{H}(\boldsymbol{x}^k)$ are, respectively, the *gradient* and *Hessian* of $F(\boldsymbol{x})$ evaluated at $\boldsymbol{x}^k$. By den Hertog [63], a solution to (8.5)–(8.6) is

$$\boldsymbol{d}^k = -\mathbf{Z}(\mathbf{Z}^{\mathrm{T}}\mathbf{H}(\boldsymbol{x}^k)\mathbf{Z})^{-1}\mathbf{Z}^{\mathrm{T}}\boldsymbol{g}(\boldsymbol{x}^k). \tag{8.7}$$

By taking suitably-sized steps α^k in the direction $\boldsymbol{d}^k$, we ensure that no $\boldsymbol{x}^{k+1} = \boldsymbol{x}^k + \alpha^k \boldsymbol{d}^k$ violates any constraint as long as the initial $\boldsymbol{x}^0$ is feasible.

For a given graph G with the set $\mathcal{S}$ of vertices and the set $E(G)$ of edges, a common approach to the Hamiltonian cycle problem is to assign a variable x_{ij} to each edge $(i, j) \in E(G)$, and solve an associated optimisation problem.

One convenient representation is to use a matrix $\mathbf{P}(\boldsymbol{x}) = [p_{ij}(\boldsymbol{x})]$, where

$$p_{ij}(\boldsymbol{x}) = \begin{cases} x_{ij} & \text{for } (i,j) \in E(G), \\ 0 & \text{otherwise.} \end{cases}$$

Arising from an embedding of the HCP in a Markov decision process, continuous optimisation problems that are equivalent to the HCP have been discovered. In particular, we demonstrated in Chapter 5 that the HCP is equivalent to solving the following optimisation problem

$$\min \{-\det(\mathbf{I} - \mathbf{P}(\boldsymbol{x}) + 1/N\boldsymbol{e}\boldsymbol{e}^{\mathrm{T}})\} \tag{8.8}$$

subject to

$$\sum_{j \in \mathcal{A}(i)} x_{ij} = 1, \quad \text{for } i \in \mathcal{S}, \tag{8.9}$$

$$\sum_{i \in \mathcal{A}(j)} x_{ij} = 1, \quad \text{for } j \in \mathcal{S}, \tag{8.10}$$

$$x_{ij} \geq 0, \quad \text{for all } (i,j) \in E(G), \tag{8.11}$$

where $\mathcal{A}(i)$ is the set of vertices reachable in one step from vertex i. We call constraints (8.9)–(8.11) the *doubly-stochastic constraints,*, and denote them by the set $\mathcal{DS}_{\boldsymbol{x}}$. Define $f(\mathbf{P}(\boldsymbol{x})) = -\det(\mathbf{I} - \mathbf{P}(\boldsymbol{x}) + 1/N\boldsymbol{e}\boldsymbol{e}^{\mathrm{T}})$, then the minimisation problem (8.8)–(8.11) can be represented as follows

$$\begin{aligned} &\min f(\mathbf{P}(\boldsymbol{x})) \\ \text{subject to}& \\ &\boldsymbol{x} \in \mathcal{DS}_{\boldsymbol{x}}. \end{aligned} \tag{8.12}$$

In this section, we solve the optimization problem (8.12) using an *interior point method* that takes advantage of the particular structure of the determinant objective function and the doubly-stochastic constraints, in order to compute the derivatives efficiently. Interior point methods are now standard and described in detail in many books (see, for example, Nocedal and Wright [82] and den Hertog [63]). Hence, we outline only the basic steps that are essential to follow our adaptation of such a method to this particular formulation of the HCP.

We now describe the *Determinant-based Interior Point* algorithm (DIPA), which should properly be regarded as a heuristic for solving the HCP. Recall that the HCP is equivalent to the optimisation problem

$$\begin{gathered} \min f(\mathbf{P}(\boldsymbol{x})) \\ \text{subject to constraints (8.9)–(8.11)} \end{gathered}$$

Constraints (8.11) are the only inequality constraints we require of $\boldsymbol{x} \in \mathcal{DS}_{\boldsymbol{x}}$. Thus, from (8.2), the auxiliary objective function is

$$F(\boldsymbol{x}) = -\det(\mathbf{I} - \mathbf{P}(\boldsymbol{x}) + 1/N\boldsymbol{e}\boldsymbol{e}^{\mathrm{T}}) - \mu \sum_{(i,j)\in E} \ln(x_{ij}). \tag{8.13}$$

We take advantage of the special structure of our formulation (the $\mathcal{DS}_{\boldsymbol{x}}$ constraints and the determinant function) to develop a specific implementation of the interior point method. After determining an initial point and initial parameters, at each iteration of DIPA we perform the following steps:

Step 1. Calculate $F(\boldsymbol{x}^k)$, and its gradient $\boldsymbol{g}(\boldsymbol{x}^k)$ and Hessian $\mathbf{H}(\boldsymbol{x}^k)$ at $\boldsymbol{x}^k$.

Step 2. Calculate the reduced gradient $\mathbf{Z}^{\mathrm{T}}\boldsymbol{g}(\boldsymbol{x}^k)$ and reduced Hessian $\mathbf{Z}^{\mathrm{T}}\mathbf{H}(\boldsymbol{x}^k)\mathbf{Z}$.

Step 3. Find a direction vector $\boldsymbol{d}^k$ and a step size α^k to determine the new point $\boldsymbol{x}^{k+1} = \boldsymbol{x}^k + \alpha^k \boldsymbol{d}^k$.

Step 4. If a variable x_{ij} has converged very close to 1 or 0, fix it to this value and alter the values of the remaining variables slightly to retain feasibility. Note that fixing a value to 1 implies others must be fixed to 0, a process we call *deflation*.

Step 5. Check if $\boldsymbol{x}^{k+1}$ corresponds to a Hamiltonian cycle after rounding the variables to 1 or 0. If so, stop and return the Hamiltonian cycle. Otherwise, repeat Steps 1–5 again.

We perform each step by invoking several component algorithms, which are described in detail in Haythorpe [62]. Heyman [64] proves that the algorithm is facilitated by the fact that for $\mathbf{P} \in \mathrm{Int}(\mathcal{DS})$, we can find the LU decomposition of $\mathbf{LU} = \mathbf{I} - \mathbf{P}$ without requiring prior permutations. Then,

$$\begin{aligned} \mathbf{W}(\mathbf{P}) &= \mathbf{I} - \mathbf{P} + 1/N\boldsymbol{e}\boldsymbol{e}^{\mathrm{T}} \\ &= \mathbf{L}(\mathbf{I} + \boldsymbol{v}\boldsymbol{w}^{\mathrm{T}})\overline{\mathbf{U}}, \end{aligned}$$

where $\mathbf{L}\boldsymbol{v} = \boldsymbol{e}$, $\overline{\mathbf{U}} = \mathbf{U} + \boldsymbol{v}\boldsymbol{e}_N^{\mathrm{T}}$, and $\overline{\mathbf{U}}^{\mathrm{T}}\boldsymbol{w} = 1/N\boldsymbol{e} - \boldsymbol{e}_N$. It can also be shown that the elements of the gradient vector and Hessian matrix of $f(\mathbf{P}(\boldsymbol{x}))$ are given by

$$\begin{aligned} g_{ij}(\mathbf{P}) &= \frac{\partial f(\mathbf{P})}{\partial x_{ij}} = -f(\mathbf{P})(\mathbf{a}_j^{\mathrm{T}}\mathbf{Q}\boldsymbol{b}_i), \\ H_{[ij],[k\ell]}(\mathbf{P}) &= \frac{\partial^2 f(\mathbf{P})}{\partial x_{ij}\partial x_{k\ell}} = g_{kj}\mathbf{a}_\ell^{\mathrm{T}}\mathbf{Q}\boldsymbol{b}_i - g_{ij}\mathbf{a}_\ell^{\mathrm{T}}\mathbf{Q}\boldsymbol{b}_k, \end{aligned}$$

where $\overline{\mathbf{U}}^{\mathrm{T}}\mathbf{a}_j = \boldsymbol{e}_j$, $\mathbf{L}\boldsymbol{b}_i = \boldsymbol{e}_i$, and $\mathbf{Q} = \mathbf{I} - \boldsymbol{v}\boldsymbol{w}^{\mathrm{T}}$.

We implemented DIPA in MATLAB and tested several sets of Hamiltonian graphs. The test results are outlined in Table 8.1. Each test set contains 50 randomly-generated Hamiltonian graphs of a certain order where each vertex degree is between 3 and 5. For each test set, we give the number of graphs (out of the 50 generated) in which the algorithm succeeds in finding a Hamiltonian cycle, the average number of iterations performed, the average number of deflations performed, the average number of function evaluations required over the course of execution, and the average running time for each graph. Since we implemented the interior point method in MATLAB, the running times are not competitive when compared to similar models implemented in compiled language. Nevertheless, we provide the running times here to demonstrate how they grow as the order N increases.

Table 8.1: Results obtained by DIPA over sets of 50 Hamiltonian graphs of order N

N	**Number solved**	**Average iterations**	**Average deflations**	**Average evaluations**	**Average running time (secs)**
20	48	20.42	9.5	20.76	1.55
40	40	86.98	27.8	87.08	12.05
60	30	198.72	42.62	201.32	54.77
80	33	372.76	65.04	372.84	196.26

Example 8.1 *We run DIPA on a 14-vertex cubic graph given by the following adjacency matrix*

$$\mathbf{A} = \begin{bmatrix} \cdot & 1 & 1 & 1 & \cdot & \cdot & \cdot & \cdot & \cdot & \cdot & \cdot & \cdot & \cdot & \cdot \\ 1 & \cdot & 1 & 1 & \cdot & \cdot & \cdot & \cdot & \cdot & \cdot & \cdot & \cdot & \cdot & \cdot \\ 1 & 1 & \cdot & \cdot & 1 & \cdot & \cdot & \cdot & \cdot & \cdot & \cdot & \cdot & \cdot & \cdot \\ 1 & 1 & \cdot & \cdot & \cdot & 1 & \cdot & \cdot & \cdot & \cdot & \cdot & \cdot & \cdot & \cdot \\ \cdot & \cdot & 1 & \cdot & \cdot & \cdot & 1 & 1 & \cdot & \cdot & \cdot & \cdot & \cdot & \cdot \\ \cdot & \cdot & \cdot & 1 & \cdot & \cdot & \cdot & \cdot & 1 & 1 & \cdot & \cdot & \cdot & \cdot \\ \cdot & \cdot & \cdot & \cdot & 1 & \cdot & \cdot & \cdot & 1 & \cdot & 1 & \cdot & \cdot & \cdot \\ \cdot & \cdot & \cdot & \cdot & 1 & \cdot & \cdot & \cdot & \cdot & \cdot & 1 & 1 & \cdot & \cdot \\ \cdot & \cdot & \cdot & \cdot & \cdot & 1 & 1 & \cdot & \cdot & \cdot & \cdot & \cdot & 1 & \cdot \\ \cdot & \cdot & \cdot & \cdot & \cdot & 1 & \cdot & \cdot & \cdot & \cdot & \cdot & 1 & 1 & \cdot \\ \cdot & \cdot & \cdot & \cdot & \cdot & \cdot & 1 & 1 & \cdot & \cdot & \cdot & \cdot & \cdot & 1 \\ \cdot & \cdot & \cdot & \cdot & \cdot & \cdot & \cdot & 1 & \cdot & 1 & \cdot & \cdot & \cdot & 1 \\ \cdot & \cdot & \cdot & \cdot & \cdot & \cdot & \cdot & \cdot & 1 & 1 & \cdot & \cdot & \cdot & 1 \\ \cdot & \cdot & \cdot & \cdot & \cdot & \cdot & \cdot & \cdot & \cdot & \cdot & 1 & 1 & 1 & \cdot \end{bmatrix}.$$

A Hamiltonian cycle is found after 8 iterations. The probability given to each edge is displayed in Figures 8.1–8.8. At iteration 1, $\mathbf{P}(\boldsymbol{x})$ *has equal probabil-*

ities to all 42 edges, but at iteration 8, one edge from each vertex contains most of the probability mass. The rounding process at the completion of iteration 8 gives a Hamiltonian cycle $1 \to 4 \to 6 \to 10 \to 12 \to 14 \to 13 \to 9 \to 7 \to 11 \to 8 \to 5 \to 3 \to 2 \to 1$.

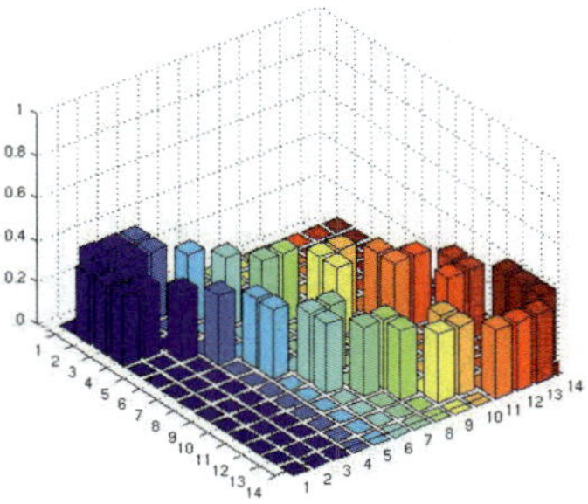

Fig. 8.1: Iteration 1

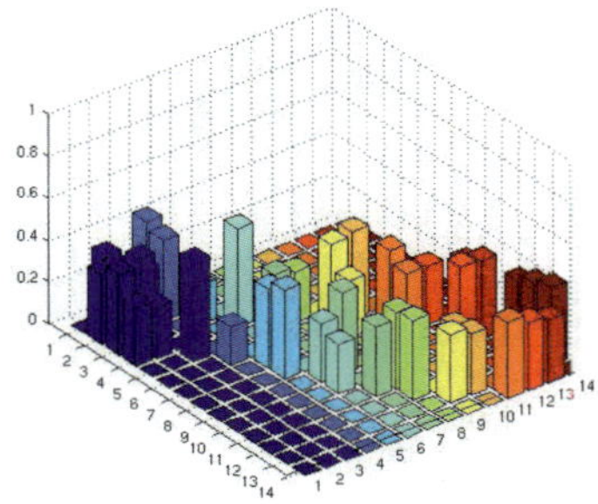

Fig. 8.2: Iteration 2

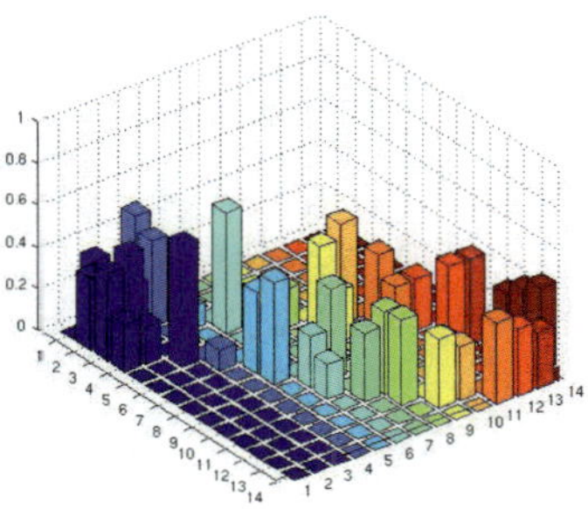

Fig. 8.3: Iteration 3

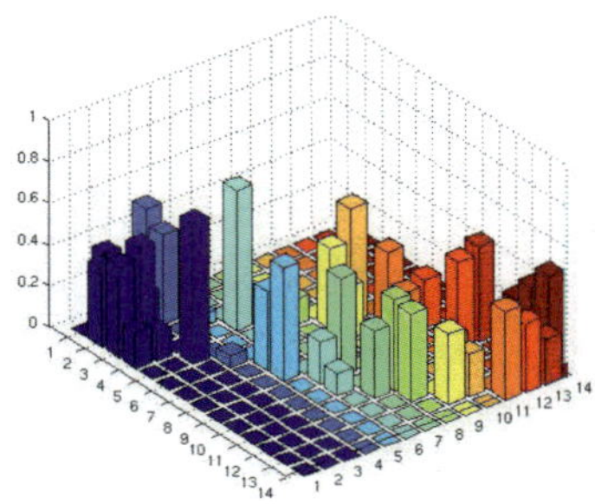

Fig. 8.4: Iteration 4

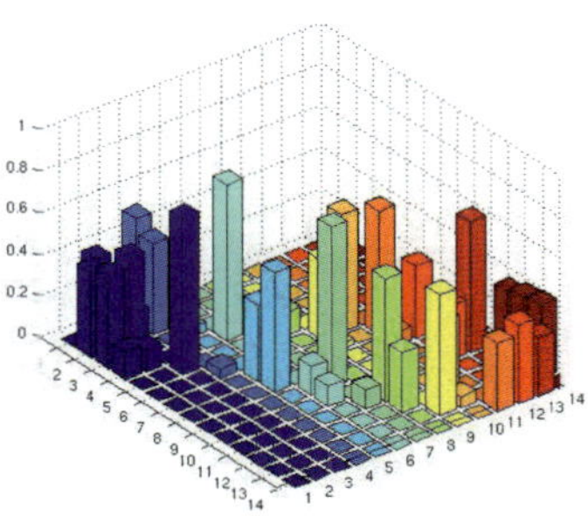

Fig. 8.5: Iteration 5

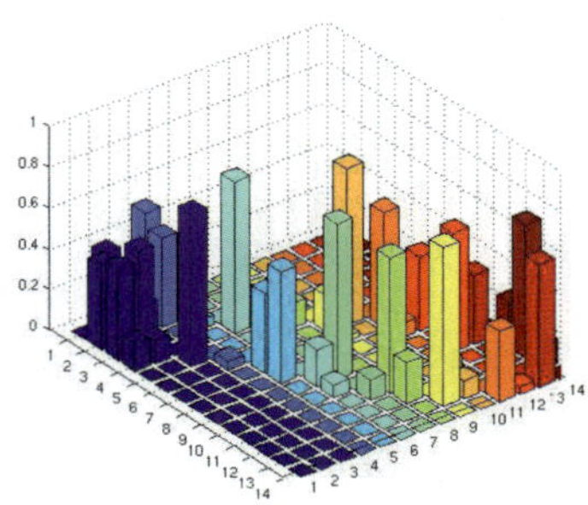

Fig. 8.6: Iteration 6

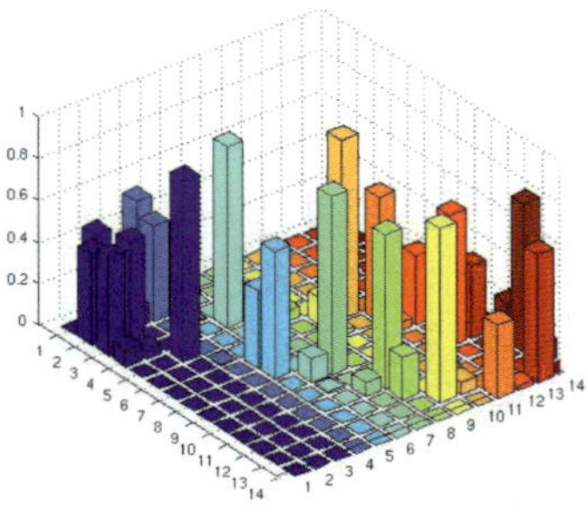

Fig. 8.7: Iteration 7

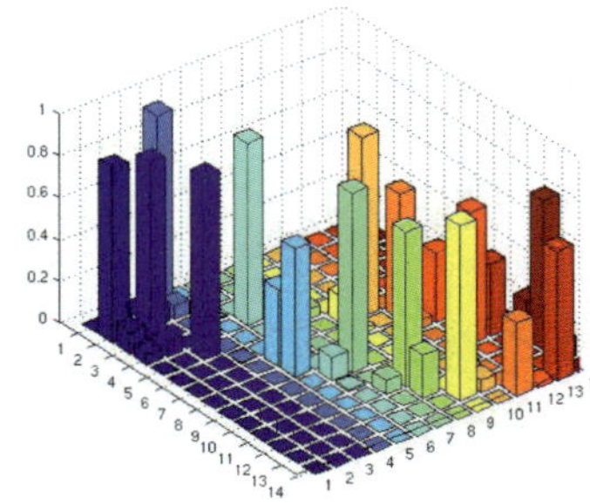

Fig. 8.8: Iteration 8

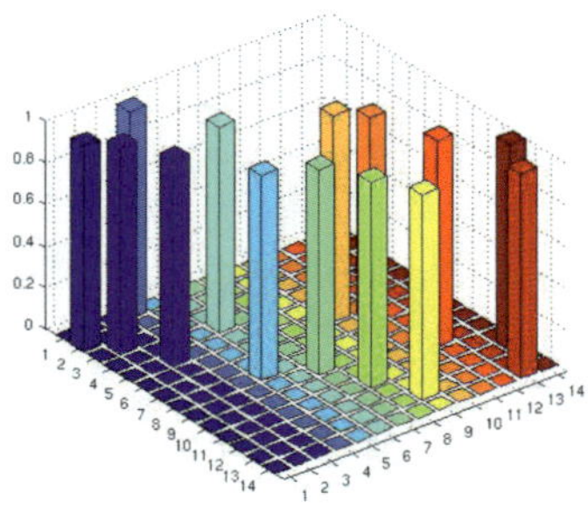

Fig. 8.9: Final Hamiltonian cycle

8.3 Cross-Entropy Algorithm

In this section, we outline a hybrid algorithm for the Hamiltonian cycle problem that is a combination of two lines of research. The first is the *Cross-Entropy method* pioneered by R. Y. Rubinstein since 1990s [94] and the second is an analytical approach based on Markov decision processes and discussed, extensively, in this book (see, in particular, Chapters 4 and 7).

The Cross-Entropy method when applied to the travelling salesman problem relies on generating random samples of tours and then constructing a sequence of transition probability matrices whose entries will, ultimately, be concentrated on the edges of an optimal tour. Of course, the Hamiltonian cycle problem can be regarded as a special case of the travelling salesman problem and is thus, in principle, solvable by this method. On the other hand, the MDP-based method too searches for a probability transition matrix entries of which are concentrated on the edges of a Hamiltonian cycle; however, it does so by solving a global optimisation problem, for instance, in the associated subspace of discounted occupational measures. The approach proposed in this section is based on Eshragh *et al.* [43] and consists of two parts:

- *MDP-part*—a static optimisation problem derived from the aforementioned embedding of a graph in a Markov decision process,
- *CE-part*—extracting information from a random sample in a manner consistent with the Cross-Entropy method. This part may be used either separately from or in conjunction with an appropriate optimisation algorithm.

Consider a linear program $\mathrm{LP}_{\mathcal{H}_\beta}$ of the following form

$$\max \sum_{i=1}^{N} \sum_{a \in \mathcal{A}(i)} \lambda_{ia} \, x_{ia}$$

subject to

$$\sum_{a \in \mathcal{A}(1)} x_{1a} - \beta \sum_{b \in \mathcal{B}(1)} x_{b1} = 1 - \beta^N, \tag{8.14}$$

$$\sum_{a \in \mathcal{A}(i)} x_{ia} - \beta \sum_{b \in \mathcal{B}(i)} x_{bi} = 0, \quad \text{for } i \in \mathcal{S}\backslash\{1\}, \tag{8.15}$$

$$\sum_{a \in \mathcal{A}(1)} x_{1a} = 1, \tag{8.16}$$

$$x_{ia} \geq 0, \quad \text{for } i \in \mathcal{S}, a \in \mathcal{A}(i), \tag{8.17}$$

$$\beta^{N-1} \leq \sum_{a \in \mathcal{A}(i)} x_{ia} \leq \beta, \quad \text{for } i \in \mathcal{S}\backslash\{1\}, \tag{8.18}$$

where coefficients λ_{ia}'s are yet to be specified. However, we note that constraints (8.14)–(8.18) correspond precisely to the constraints (7.21)–(7.24) of the Wedged-MIP heuristics. A vector $\boldsymbol{x}$ satisfying (8.14)–(8.18) is said to be *quasi-Hamiltonian* if for each $i \in \mathcal{S}$, there exists unique $a_i \in \mathcal{A}(i)$ such that $a_i \in \operatorname{argmax}\{x_{ia} | a \in \mathcal{A}(i)\}$ and $(a_1, \ldots, a_N)$ identifies a Hamiltonian cycle in the underlying graph. We denote by $\mathcal{H}_\beta$ the polytope defined by constraints (8.14)–(8.17), and by $\mathcal{WH}_\beta$ the polytope defined by constraints (8.14)–(8.18).

In this section, we develop a new hybrid algorithm, Hybrid Simulation-Optimisation (HSO), for the HCP by synthesizing Cross-Entropy (CE) and discounted Markov decision processes. For an excellent introduction to the Cross-Entropy method, we refer to Rubinstein and Kroese [95]. We note that for any Hamiltonian cycle in the graph G, there exists some objective function coefficient matrix $\boldsymbol{\Lambda} = [\lambda_{ij}]$ for which the corresponding Hamiltonian solution of (8.14)–(8.18) maximises that objective function. Thus, we try to construct such a matrix by applying an adaptive procedure as in the CE method, and then supply the matrix to $\mathrm{LP}_{\mathcal{H}_\beta}$. We use the terms *Hamiltonian cycle* and *tour* interchangeably.

Consider a given graph G which may or may not contain Hamiltonian cycles.

Preliminary Set-up. Augment G with *artificial edge* to make the given graph a complete one with no self-loops. The edges in the original graph G are now called *authentic edges* to differentiate them from the newly-inserted artificial edges. Assign a length ℓ_{ia} to each edge (i, a) as follows:

$$\ell_{ia} \sim \begin{cases} U(0, \omega) & \text{if } (i, a) \text{ is an authentic edge,} \\ U(\mu, \mu + \sigma) & \text{otherwise,} \end{cases}$$

where $U(a, b)$ denotes the uniform distribution on the interval $[a, b]$. Here, ω, μ, and σ are arbitrary positive real numbers, the only restriction is that $\omega \ll \mu$, where $\ll$ (respectively, $\gg$) means *sufficiently smaller* (respectively, *sufficiently greater*). We can, for example, set $\omega = O(N^{1/\eta})$, $\mu = O(N^{1+\varsigma/\eta})$, and $\sigma = \omega$, for $\eta, \varsigma > 1$. Since lengths are generated from continuous distributions, the *total length* $L(\Theta)$ of a random tour Θ, defined to be the finite sum of N continuous random variables, is a continuous random variable. Hence, for two distinct tours Θ_i and Θ_j,

$$P\{L(\Theta_i) = L(\Theta_j)\} = 0.$$

This means that, with probability 1, the shortest tour is unique and if the original graph is Hamiltonian, then the length of an optimal tour must be less than $N\omega$. If the graph is not Hamiltonian, that length will be at least μ.

Set up the initial transition probability matrix $\mathbf{P}_0 = [p_{0,ia}]$, where

$$p_{0,ia} = \begin{cases} \dfrac{1}{|\mathcal{A}(i)|} - \dfrac{N - 1 - |\mathcal{A}(i)|}{|\mathcal{A}(i)|}\varepsilon & \text{if } (i, a) \text{ is an authentic edge,} \\ \varepsilon & \text{otherwise.} \end{cases}$$

Here, $0 < \varepsilon < 1/(N - 1 - |\mathcal{A}(i)|)$ is a small positive real number. We can interpret the element $p_{t,ia}$ of a corresponding matrix $\mathbf{P}_t$ as the probability of (i, a) belonging to the shortest tour, given the observations obtained up to the end of iteration number t.

Step 0. Set $t = 0$.

Step 1. Set $t = t + 1$.

Step 2. Obtain random tours $\Theta_1, \Theta_2, \ldots, \Theta_n$ from $\mathbf{P}_{t-1}$ as follows:

- generate 100ζ percent of the tours based on the rows of $\mathbf{P}_{t-1}$, and
- generate the remaining $100(1-\zeta)$ percent based on the columns of $\mathbf{P}_{t-1}$,

where $\zeta \in [0,1]$ is chosen arbitrarily. If $\zeta = 1$, the random tour generation is the same as in CE. The former tours are called *direct tours*, the latter are called *reverse tours*[1].

If a tour with total length of less than $N\omega$ has been generated,
go to Step 6;
else, go to Step 3.

Step 3. Update matrix $\mathbf{P}_t$ with the best m generated tours, $0 < m \leq N$, in an analogous way to that mentioned in CE algorithm (Rubinstein and Kroese [95]), that is,

$$\gamma_t^* \in \arg\min_{\gamma}\{1/m \sum_{j=1}^{m} e^{-L(\Theta_{(j)})/\gamma} \geq \rho\},$$

$$p_{t,ia}^* = \frac{\displaystyle\sum_{1 \leq j \leq m : \Theta_{(j)} \ni (i,a)} e^{-L(\Theta_{(j)})/\gamma_t^*}}{\displaystyle\sum_{j=1}^{m} e^{-L(\Theta_{(j)})/\gamma_t^*}},$$

where the summation over "$1 \leq j \leq m : \Theta_{(j)} \ni (i,a)$" denotes summation over the set of all tours passing through the arc (i,a) selected from the set of best m generated tours, and

$$\mathbf{P}_t = (1-\alpha)\mathbf{P}_{t-1} + \alpha\mathbf{P}_t^*,$$

where α is a smoothing parameter chosen from the interval (0,1).

If $\{t$ is large enough$\}$ AND $\{(p_{t,ia} \gg 0$ for some $i \in \mathcal{S}$ and $a \notin \mathcal{A}(i))$
or $(p_{t,ia} \ll 1$ for all $i \in \mathcal{S}$ and $a \in \mathcal{A}(i))\}$,
go to Step 7;
else, go to Step 4.

Step 4. Define $\widetilde{\boldsymbol{\Lambda}} = \mathbf{P}_t$. Set all entries of the matrix $\widetilde{\boldsymbol{\Lambda}}$ corresponding to artificial edges equal to zero, then normalise each row to keep it a transition probability matrix. Denote the resulting matrix by $\boldsymbol{\Lambda} = [\lambda_{ia}]$, and consider it as the objective function coefficient matrix of $\text{LP}_{\mathcal{H}_\beta}$, then solve this linear

[1] We note that $\mathbf{P}_{t-1}$ may not be a doubly stochastic matrix. Hence, in an intermediate phase (to generate reverse tours), each column of this matrix is normalised.

program. Denote an optimal solution by $\boldsymbol{x}^* = \{x^*_{ia} | i \in \mathcal{S},\ a \in \mathcal{A}(i)\}$.

If $\boldsymbol{x}^*$ is a quasi-Hamiltonian solution,
then go to Step 6,
else go to Step 5.

Step 5. If for any $t > r$ and some r, say $r = 20$,

$$\mathcal{L}_t = \mathcal{L}_{t-1} = \cdots = \mathcal{L}_{t-r+1}, \tag{8.19}$$

reset $\mathbf{P}_t$ by $\mathbf{P}_t = (1-\hat{\alpha})\mathbf{P}_0 + \hat{\alpha}\mathbf{P}_t$, where $\hat{\alpha}$ is a fixed parameter chosen from the interval $(0, 1)$.

Go to Step 1.

Step 6. The graph is Hamiltonian, STOP.

Step 7. The graph is likely to be non-Hamiltonian, STOP.

Since it is proved in Margolin [77] that typical modifications of the CE algorithm converge to an optimal solution in a finite number of steps with probability one, it could be argued that the CE-part might be sufficient for the HSO algorithm. Of course, even the CE-part differs slightly from the basic CE method[2], as the concept of generating *reverse tours* is introduced in the above HSO algorithm. However, the main reason for augmenting the MDP-part is the numerical performance. Our numerical experiments show that the MDP-part has a significant effect on decreasing the search-time for finding a tour in a Hamiltonian graph.

We implement the HSO algorithm in MATLAB using an efficient random number generator taken from Gentle [55, Section 1.2]. For solving $\mathrm{LP}_{\mathcal{H}_\beta}$, we use CPLEX software. Furthermore, in order to prevent the algorithm from jamming, if after some consecutive iterations, say 20, the best solution achieved, $\mathcal{L}_t$, does not change, then the probability transition matrix $\mathbf{P}_t$ is reset as expressed in the Step 5 of the HSO algorithm. In all numerical examples, we set $\omega = N$, $\mu = N^3$, $\sigma = N^2$, $\varepsilon = 1/N^5$, $m = 0.85N$, $\rho = 1/N^3$, $\beta = 0.9999$ and $\hat{\alpha} = 0.25$.

Preliminary numerical results show that, in most of Hamiltonian graphs, the basic CE method does not converge to a deterministic policy corresponding to a Hamiltonian cycle containing only authentic edges. More precisely, in these Hamiltonian cycles, most of early edges (starting at home vertex 1) are authentic and some of the latter edges are artificial. This is understandable,

[2] By the *basic CE method*, we mean that all generated tours are direct tours, that is, the CE-part with $\zeta = 1$.

as authentic edges have, potentially, larger probabilities, they have higher chances of being chosen when the algorithm tries to generate a random tour. This means that most of authentic edges are generated in early steps, and then, after the algorithm reaches some vertex, there are no (or very few) authentic edges available and hence some artificial edge is selected.

To overcome this difficulty, we generate reverse tours. Numerical results show that generating some reverse tours as a part of a sample can, significantly, speed up the rate of convergence to a Hamiltonian cycle comprising of all authentic edges. For example, we present the results of solving the 8×8 Knight's Tour problem with different values of ζ in Table 8.2. In these examples, $N = 1000$ and $\alpha = 0.8$. The first column of Table 8.2 contains 11 distinct values of ζ, the second column shows the number of iterations that the algorithm took to find a Hamiltonian cycle containing only authentic edges and the last column determines which part of the algorithm caused termination.

Table 8.2: Solving the HCP for the 8×8 Knight's Tour problem.

ζ	**Iteration**	**Terminated by**
0.50	24	CE-part
0.55	19	MDP-part
0.60	6	MDP-part
0.65	5	MDP-part
0.70	36	CE-part
0.75	17	CE-part
0.80	18	CE-part
0.85	19	CE-part
0.90	19	CE-part
0.95	15	CE-part
1.00	19	CE-part

Table 8.2 indicates that the value of $\zeta = 0.65$ seems to be the most appropriate selection for this parameter. Interestingly, this value of ζ led to the best solution times in many other numerical examples as well. Thus, we set $\zeta = 0.65$ in all subsequent examples.

With these parameter settings, we solved many Hamiltonian graphs with the HSO algorithm, in the range of 6–256 vertices. In all these examples, the HSO algorithm found a Hamiltonian cycle containing only authentic edges in the graph. Table 8.3 demonstrates results of several such examples. The first column of this table contains identification of each graph, the second one is the sample size, the third and fourth ones are the same as columns two and three in Table 8.2. As exhibited in Table 8.3, when the order of a graph is small—say, fewer than 50 vertices—there is a high chance to generate a tour in the original graph at random and, consequently, the algorithm is terminated in the CE-part. However, when the order of a graph is moderate to large—say, more than 50 vertices—the algorithm is more likely to terminate by finding a Hamiltonian cycle through the MDP-part. Moreover, while Rubinstein and Kroese [95, Section 4.7] propose that the sample size should be order of $O(N^2)$ to obtain good results from the basic CE method, here, by generating reverse tours as well as solving $LP_{\mathcal{H}_\beta}$, this order reduces to $O(N)$.

Table 8.3: Solving the HCP for various graphs

Graph	n	**Iteration**	**Terminated by**
Envelope Graph	20	1	CE-part
Hamiltonian Graph, N = 10	20	1	CE-part
Hamiltonian Graph, N = 12	20	1	CE-part
Hamiltonian Graph, N = 18	40	2	CE-part
Hamiltonian Graph, N = 20	50	2	CE-part
Hamiltonian Graph, N = 24	50	1	CE-part
Hamiltonian Graph, N = 30	100	4	CE-part
Hamiltonian Graph, N = 38	100	2	CE-part
Hamiltonian Graph, N = 48	400	9	MDP-part
8×8 Knight's Tour	1000	5	MDP-part
10×10 Knight's Tour	1000	47	MDP-part
12×12 Knight's Tour	2000	42	MDP-part
14×14 Knight's Tour	3000	397	MDP-part
16×16 Knight's Tour	4000	277	MDP-part

Another advantage of this algorithm is that including both CE-part and MDP-part can alleviate drawbacks of each part when considered separately. More precisely, when there are plenty of edges in the graph, there will be a high chance to generate a Hamiltonian cycle in the graph, at random, and while there is a small number of edges in the graph (for instance, in the family of cubic graphs) the MDP-part helps to find a Hamiltonian cycle.

8.4 Open Algorithmic Problems

In the last two chapters, we described a number of promising algorithmic developments that exploit the stochastic approach. The DIPA algorithm described earlier in Section 8.2 is based on searching the domain $\mathcal{DS}$ of policies inducing doubly stochastic probability transition matrices. It employs an interior point method that aimed to approach, through the interior, a global optimum of the problem

$$\max_{\mathbf{P}\in\mathcal{DS}} \det \mathbf{W}(\mathbf{P}), \tag{DM}$$

where $\mathbf{W}(\mathbf{P}) = \mathbf{I} - \mathbf{P} + 1/N\mathbf{J}$. Important special properties that made this a tractable optimisation problem include: (a) simplicity of the structure of the null space of the doubly stochastic constraints, and (b) availability of an LU-type decomposition of the matrix $\mathbf{W}(\mathbf{P})$ when $\mathbf{P}$ lies in the interior of $\mathcal{DS}$. Unfortunately, $\det \mathbf{W}(\mathbf{P})$ is a high-order polynomial in the decision variables that are the entries of $\mathbf{P}$.

However, an entirely novel re-formulation of (DM) in terms of the variables of the above LU-decomposition should be possible. Namely, the decision variables would be the entries of the matrices $\mathbf{L}$ and $\mathbf{U}$ rather than of $\mathbf{P}$. The triangular structure of the $\mathbf{L}$ and $\mathbf{U}$ matrices in such a re-formulation should significantly reduce the order of the polynomial of an objective function to be optimised. The design and implementation of an effective algorithm based on this notion constitutes a worthwhile, still open, problem.

A fundamental algorithmic difficulty of the HCP is the extreme "rarity" of Hamiltonian cycles among the spanning subgraphs of some (typically sparse) Hamiltonian graphs. Consequently, it is natural to try to develop methods where the search for these cycles is over domains where they are not as rare.

Indeed, we see that the algorithm described in Section 8.3 combines the modern simulation technique known as the Cross-Entropy method with the polyhedral methods based on the space $\mathcal{X}_\beta$ of discounted occupational measures discussed in Chapter 4. The idea is to use the importance sampling type approach at initial iterations of the algorithm and then switch to one of the classical optimisation techniques. In this approach, elite sub-samples steer the method into the regions of the feasible domain where finding a Hamiltonian cycle may be more likely. However, it is also natural to consider the question whether the domain $\mathcal{X}_\beta$ can be refined so that Hamiltonian cycles are no longer so rare in the family of extreme points of the refined domain.

Exploiting the development in Chapters 4 and 7, we can constrain $\mathcal{X}_\beta$ by a very special single cut and a normalisation to construct a polytope $\widetilde{\mathcal{X}}_\beta$ that is nonempty whenever the graph is Hamiltonian and which, in a prescribed sense, contains all the Hamiltonian cycles among its extreme points. While the latter polytope has already been exploited as a base of the Wedged-MIP heuristic, there is some evidence that it may also reveal certain fundamental properties of Hamiltonian graphs, when the parameter β is sufficiently close to 1.

Now, transform the polytope $\mathcal{X}_\beta$ defined by constraints (4.18)–(4.20) by setting $\mu = 0$, multiplying all constraints by $1 - \beta^N$ and changing variables to $\bar{x}_{ia} = (1-\beta^N)x_{ia}$, for all $i \in \mathcal{S}, a \in \mathcal{A}(i)$. Let us denote the transformed polytope by $\mathcal{H}_\beta$. The results of Sinclair [97] naturally lead to the consideration of random walks on extreme points of specially constructed subsets of $\mathcal{H}_\beta$. In fact, in Eshragh and Filar [42] two smaller polytopes $\mathcal{WH}_\beta$ and $\mathcal{WHR}_\beta^p$ are constructed in such a way that

$$\mathcal{WHR}_\beta^p \subset \mathcal{WH}_\beta \subset \mathcal{H}_\beta, \tag{8.20}$$

and, more importantly, the only extreme points that $\mathcal{WHR}_\beta^p$ and $\mathcal{H}_\beta$ have in common correspond precisely to Hamiltonian cycles. This is, schematically, portrayed in Figure 8.10. Thus, the distance between extreme points of $\mathcal{WHR}_\beta^p$ and $\mathcal{H}_\beta$ is mimimised to zero at, precisely, the points corresponding to Hamiltonian cycles.

Then, by borrowing probability transition adaptation ideas from the Cross-Entropy method (see Rubinstein and Kroese [95]), Eshragh and Filar [42] construct a random walk pivoting algorithm on extreme points of $\mathcal{WHR}_\beta^p$ that converges surprisingly fast in test problems. For instance, increasing the order of a Hamiltonian graph from 20 to 100 increases the number of random pivots required to find a Hamiltonian cycle from 10 to only 29.

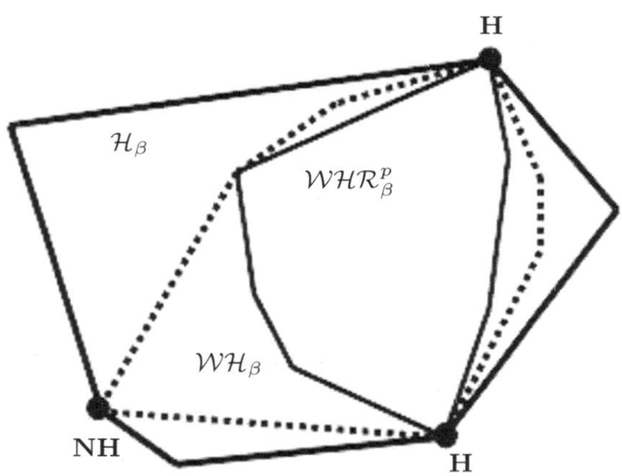

Fig. 8.10: Nesting of occupational measures polytopes.

Eshragh and Filar [42] prove that this random walk algorithm will find a Hamiltonian cycle (if there is one) in finite time, with probability one. However, what makes this algorithm more interesting is, that the observed slow growth of the number of random pivots in the graph order N suggests that the following, much more powerful, result may also be true.

Conjecture 8.1. Let $\mathcal{Y}$ be a suitably-designed random walk algorithm and Y be the random variable denoting the number of extreme points that $\mathcal{Y}$ will need to examine to identify a Hamiltonian cycle, if there is one. Then there exist polynomials $q_1(\log(\beta), N)$ and $q_2(\log(\beta), N)$ (with positive coefficients of the leading power of N) such that for β sufficiently near 1,

$$P\{Y > q_1(\log(\beta), N)\} \leq e^{-q_2(\log(\beta), N)}.$$

We conclude this section with some additional, speculative, comments. In particular, it can be argued that the stochastic perspective of the Hamiltonian cycle problem opens new opportunities for advancing the theory of detecting special structures in complex graphs.

Once we think of a graph as a stochastic control system with the controller making randomised choices at each vertex (according to a policy $\boldsymbol{f}$), we can also think of a graph as a "transmitter" generating signals that are influenced both by the policy and by the underlying, discrete, structure of the graph. This opens prospects of detecting, possibly very rare structures (like Hamiltonian cycles) in the graph by generating prescribed signals, and analysing their properties in response to changes in controls. The rich subjects of Markov processes, stochastic control, simulation and statistics offer many interesting tools for this kind of investigations.

Part V
Geometric Approaches

Chapter 9
Self-similar Structure and Hamiltonicity

9.1 Introduction

The class of cubic graphs provides a convenient laboratory for studying Hamiltonicity. This is because the Hamiltonian cycle problem is already NP-complete for this class [53], and because there is freely available and reliable software for enumerating all connected cubic graphs with N vertices (see, for example, Meringer [79]). This offers an opportunity to study the whole populations of these graphs with the goal of understanding the special nature of the sub-populations of non-Hamiltonian graphs.

In this chapter, we exhibit a characteristic—self-similarity—structure of the class of all cubic graphs that stems from the spectra of their adjacency matrices. The structure has a fractal threadlike appearance. Points with coordinates given by the mean and variance of certain simple functions of graph eigenvalues cluster around line segments that we call *filars*. These are identified by the number of triangles in graphs. Zooming in reveals that these clusters split into smaller segments, *sub-filar*, identified by the number of quadrangles in graphs, etc. Collectively, we refer to this fractal (or self-similar) structure, discovered in numerical experiments, as *multifilar structure*. We also provide a mathematical explanation of this phenomenon based on the Ihara-Selberg trace formula, and compute the coordinates and slopes of all filars in terms of Bessel functions of the first kind. We also consider the location of non-Hamiltonian cubic graphs in the multifilar structure and observe that cubic bridge graphs not only take up characteristic positions in that structure but also constitute a great majority of all non-Hamiltonian cubic graphs. This observation leads to the, still open, conjecture about the prevalence of cubic bridge graphs, discussed in Section 9.4. We note that the analysis presented here for cubic graphs can be extended to regular graphs of degree d.

9.2 Preliminaries

The famous Selberg trace formula relates the eigenvalue spectrum of the Laplace operator on a hyperbolic surface to its length spectrum—the collection of lengths of closed geodesics counted with multiplicities. An immediate consequence of the Selberg trace formula is that the eigenvalue spectrum and the length spectrum uniquely determine one another. A similar result is valid for regular graphs. Two important references for the Selberg trace formula are Ihara [68] and Ahumada [1]. To formulate this formula precisely, we need to introduce some terminology, starting with a few definitions concerning geodesics.

An *elementary homotopy* is a transformation of a closed walk of the following form

$$(v_1, v_2, \ldots, v_i, \ldots, v_{k-1}, v_k, v_1) \mapsto (v_1, v_2, \ldots, v_i, v_j, v_i, \ldots, v_{k-1}, v_k, v_1),$$

where v_j is a neighbor of v_i, and the arrow can also be pointing in the opposite direction

$$(v_1, v_2, \ldots, v_i, v_j, v_i, \ldots, v_{k-1}, v_k, v_1) \mapsto (v_1, v_2, \ldots, v_i, \ldots, v_{k-1}, v_k, v_1).$$

If one closed walk can be obtained from another by a sequence of elementary homotopies (in either direction of the arrow), they are said to be *homotopic*. A *homotopy class* is a set of closed walks such that every pair of closed walks in the set are homotopic. In a homotopy class of closed walks, the shortest walk is called a *closed geodesic*. In other words, *a closed geodesic* is a closed walk with no cycles of length 2, that is, $v_i \neq v_{i+2}$ for all i. As we are only concerned with closed geodesics in this book, we will simply refer to them as *geodesics*.

Also known as a *short geodesic*, a *contractible* is a geodesic of length 0, or equivalently, a single vertex. A homotopy class of closed walks containing a geodesic of length 0 is equivalent to a homotopy class of closed walks such that each member is either a single vertex or a union of two or more joint cycles of length 2. A *long geodesic* is a geodesic of length greater than zero, which from now on we refer to as simply a *geodesic*. A geodesic of length 3, 4 or 5 is equivalent to a cycle of length 3, 4 or 5. On the other hand, a geodesic of length 6 or longer can be a union of joint cycles. Consider a geodesic $g = (v_1, v_2, \ldots, v_\ell, v_1)$ of length ℓ. Another geodesic is said to be a k-multiple of g, denoted as g^k, if it simply traces out g for k times: $g^k = (\{v_1, v_2, \ldots, v_\ell, v_1\}, \{v_1, v_2, \ldots, v_\ell, v_1\}, \ldots, \{v_1, v_2, \ldots, v_\ell, v_1\})$. A geodesic is said to be *primitive* if it is not a multiple of a shorter geodesic.

9.3 Self-similar Multifilar Structure

We denote by λ_i, $i = 1, \ldots, N$, eigenvalues of the adjacency matrix $\mathbf{A}$ of a given cubic graph, and choose $t \in (0, 1/3)$ to guarantee that the inverse of $\mathbf{I} - t\mathbf{A}$ exists. Then, it is clear that eigenvalues of $(\mathbf{I} - t\mathbf{A})^{-1}$ are $(1 - t\lambda_i)^{-1}$. For each adjacency matrix $\mathbf{A}$, we define the expected value function of $(1 - t\lambda_i)^{-1}$, with respect to uniform distribution on the N eigenvalues, to be

$$\begin{aligned}\mu(\mathbf{A}, t) &= \frac{1}{N}\mathrm{Tr}[(\mathbf{I} - t\mathbf{A})^{-1}] \\ &= \frac{1}{N}\sum_{i=1}^{N}\frac{1}{1 - t\lambda_i},\end{aligned} \tag{9.1}$$

and the variance function to be

$$\begin{aligned}\sigma^2(\mathbf{A}, t) &= \frac{1}{N}\mathrm{Tr}[(\mathbf{I} - t\mathbf{A})^{-2}] - \mu^2(\mathbf{A}, t) \\ &= \frac{1}{N}\sum_{i=1}^{N}\frac{1}{(1 - t\lambda_i)^2} - \mu^2(\mathbf{A}, t).\end{aligned} \tag{9.2}$$

For numerical experiments presented here, we choose $t = 1/9 \in (0, 1/3)$. We illustrate this procedure with the following example.

Example 9.1 *We consider the labelled Petersen graph (Figure 9.1), which was also considered in Example 2.2.*

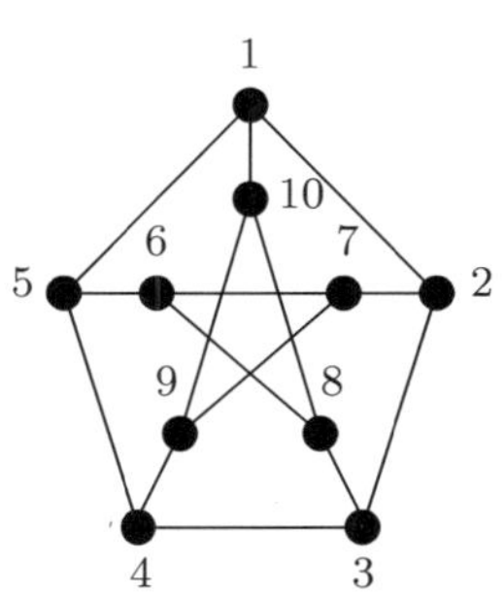

Fig. 9.1: Labelled Petersen graph

The adjacency matrix $\mathbf{A}$ *of the Petersen graph, the only 10-vertex cubic graph that is non-Hamiltonian, is*

$$\mathbf{A} = \begin{bmatrix} \cdot & 1 & \cdot & \cdot & 1 & \cdot & \cdot & \cdot & \cdot & 1 \\ 1 & \cdot & 1 & \cdot & \cdot & \cdot & 1 & \cdot & \cdot & \cdot \\ \cdot & 1 & \cdot & 1 & \cdot & \cdot & \cdot & 1 & \cdot & \cdot \\ \cdot & \cdot & 1 & \cdot & 1 & \cdot & \cdot & \cdot & 1 & \cdot \\ 1 & \cdot & \cdot & 1 & \cdot & 1 & \cdot & \cdot & \cdot & \cdot \\ \cdot & \cdot & \cdot & \cdot & 1 & \cdot & 1 & 1 & \cdot & \cdot \\ \cdot & 1 & \cdot & \cdot & \cdot & 1 & \cdot & \cdot & 1 & \cdot \\ \cdot & \cdot & 1 & \cdot & \cdot & 1 & \cdot & \cdot & \cdot & 1 \\ \cdot & \cdot & \cdot & 1 & \cdot & \cdot & 1 & \cdot & \cdot & 1 \\ 1 & \cdot & \cdot & \cdot & \cdot & \cdot & \cdot & 1 & 1 & \cdot \end{bmatrix} \tag{9.3}$$

Thus,

$$\mu(\mathbf{A}, 1/9) = 1/10\mathrm{Tr}[(\mathbf{I} - 1/9\mathbf{A})^{-1}] = 1.0398, \tag{9.4}$$

and

$$\sigma^2(\mathbf{A}, 1/9) = 1/10\mathrm{Tr}[(\mathbf{I} - 1/9\mathbf{A})^{-2}] - \mu^2(\mathbf{A}, 1/9) = 0.0445, \tag{9.5}$$

and we obtain the point (1.0398, 0.0445) *for the Petersen graph.*

Figure 9.2 shows the plot of $(\mu(\mathbf{A}), \sigma^2(\mathbf{A}))$ across all 509 cubic graphs of order 14, which exhibits a self-similar multifilar structure.

In order to explain why pairs of coordinates of certain graphs belong to particular filars, we start by establishing alternative formulae for $\mu(\mathbf{A}, t)$ and $\sigma^2(\mathbf{A}, t)$. Let p_ℓ be the number of all walks of length ℓ in the graph G, for $\ell \geq 0$. It is well known (see, for example, van Dam and Haemers [28]) that the ith diagonal element of the ℓth power of $\mathbf{A}$, namely $[\mathbf{A}^\ell]_{ii}$, is the number of closed walks of length ℓ starting at i. Hence, for any t such that the spectral radius of $t\mathbf{A}$ is less than 1,

$$\begin{aligned} \mu(\mathbf{A}, t) &= \frac{1}{N}\mathrm{Tr}\left[(\mathbf{I} - t\mathbf{A})^{-1}\right] \\ &= \frac{1}{N}\mathrm{Tr}\left[\mathbf{I} + t\mathbf{A} + t^2\mathbf{A}^2 + \cdots\right] \\ &= \frac{1}{N}\left[N + tp_1 + t^2p_2 + t^3p_3 + \cdots\right] \\ &= \frac{1}{N}\sum_{i=0}^{\infty} t^i p_i. \end{aligned} \tag{9.6}$$

The infinite sum in (9.6) resembles the trace of the matrix exponential of $t\mathbf{A}$

$$\mathrm{Tr}[e^{t\mathbf{A}}] = \sum_{i=0}^{\infty} \frac{t^i}{i!} p_i. \tag{9.7}$$

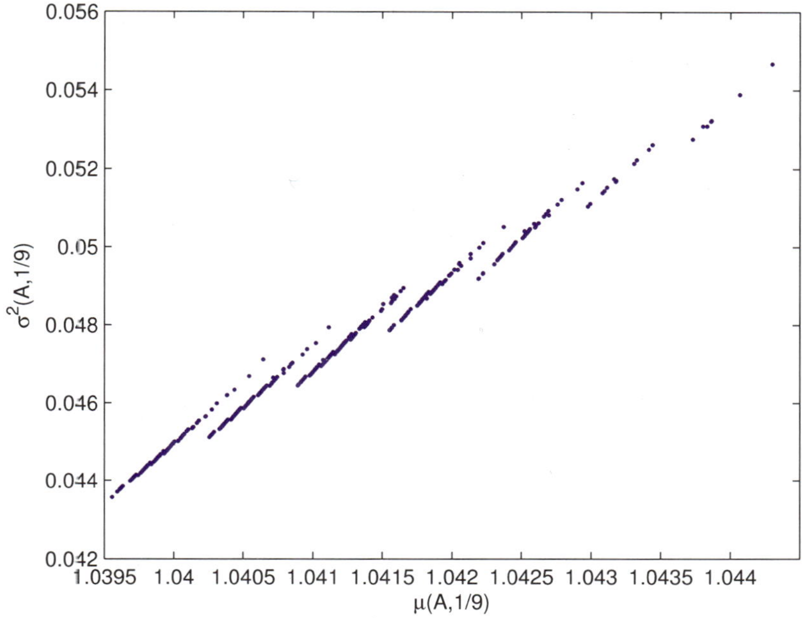

Fig. 9.2: Mean-Variance (trace of resolvent) plot for cubic graphs of order 14

Mnev [30, Lemma 1] presents a recent proof of the Selberg trace formula for regular graphs, which gives a closed-form expression for (9.7). We restate here the theorem, in our context.

Theorem 9.1. *For a given d-regular graph G of order N,*

$$\mathrm{Tr}[e^{t\mathbf{A}}] = N\frac{q+1}{2\pi}\int_{-2\sqrt{q}}^{2\sqrt{q}} ds e^{st}\frac{\sqrt{4q-s^2}}{(q+1)^2-s^2} + \sum_{\gamma\in G}\Lambda(\gamma)q^{|\gamma|/2}I_{|\gamma|}(2\sqrt{q}t), \tag{9.8}$$

where $q = d-1$, $|\gamma|$ is the length of a long geodesic γ, $\Lambda(\gamma) = |\gamma'|$ if γ is a multiple of a primitive geodesic γ', $\Lambda(\gamma) = |\gamma|$ if γ is primitive itself, and $I_{|\gamma|}(z)$ denotes the Bessel function of the first kind.

In (9.8), the first summand is the contribution of the homotopy classes of short geodesics, and the second summand is the contribution of the homotopy classes of long geodesics. Recall that a geodesic is the shortest walk of its homotopy class, and a geodesic is a short geodesic if it has length 0 and a long geodesic otherwise. Let $C_1(t)$ and $C_2(t)$ be the contributions to (9.6) from homotopy classes of short geodesics and long geodesics, respectively. Then,

$$\mu(\mathbf{A},t) = \frac{1}{N}\left(C_1(t) + C_2(t)\right). \tag{9.9}$$

By Mnev [80, equations (19), (20)], the first contribution is

$$C_1(t) = N\frac{3\sqrt{1-8t^2}-1}{2(1-9t^2)}. \tag{9.10}$$

and by Mnev [80, equations (26), (27)], the second contribution is

$$C_2(t) = \sum_{\gamma\in G}\frac{\Lambda(\gamma)}{\sqrt{1-8t^2}}\{(1-\sqrt{1-8t^2})/(4t)\}^{|\gamma|}, \tag{9.11}$$

Define $\Theta(t) = (1-\sqrt{1-8t^2})/(4t)$. Substituting (9.10) and (9.11) into (9.9), we have

$$\mu(\mathbf{A},t) = \frac{3\sqrt{1-8t^2}-1}{2(1-9t^2)} + \frac{1}{N}\sum_{\gamma\in G}\frac{\Lambda(\gamma)}{\sqrt{1-8t^2}}\Theta^{|\gamma|}(t). \tag{9.12}$$

Recall that a geodesic is said to be a k-multiple of g, denoted as g^k, if it traces out g exactly k times, namely,

$$g^k = (\{v_1,v_2,\ldots,v_\ell,v_1\},\{v_1,v_2,\ldots,v_\ell,v_1\},\ldots,\{v_1,v_2,\ldots,v_\ell,v_1\}).$$

Consequently, if we consider some geodesic γ that is a k-multiple of a primitive geodesic γ', then $|\gamma| = k|\gamma'|$. The set of all long geodesics in a graph G can be partitioned into sets of primitive geodesics and their k-multiples, for $k = 1,\ldots,N$. Therefore, by denoting primitive long geodesics by ζ, we transform equation (9.12) into

$$\begin{aligned}\mu(\mathbf{A},t) &= \frac{3\sqrt{1-8t^2}-1}{2(1-9t^2)} + \frac{1}{N}\sum_{\zeta\in G}\sum_{k=1}^{\infty}\frac{|\zeta|}{\sqrt{1-8t^2}}\Theta^{k|\zeta|}(t)\\ &= \frac{3\sqrt{1-8t^2}-1}{2(1-9t^2)} + \frac{1}{N}\sum_{\zeta\in G}\frac{|\zeta|}{\sqrt{1-8t^2}}\sum_{k=1}^{\infty}\Theta^{k|\zeta|}(t).\end{aligned}$$

When $t\in(0,1/3)$, we have that $\Theta(t)\in(0,1)$. Hence,

$$\begin{aligned}\mu(\mathbf{A},t) &= \frac{3\sqrt{1-8t^2}-1}{2(1-9t^2)} + \frac{1}{N}\sum_{\zeta\in G}\frac{|\zeta|}{\sqrt{1-8t^2}}\frac{\Theta^{|\zeta|}(t)}{1-\Theta^{|\zeta|}(t)}\\ &= \frac{3\sqrt{1-8t^2}-1}{2(1-9t^2)} + \frac{1}{N}\sum_{\zeta\in G}\frac{|\zeta|}{\sqrt{1-8t^2}}\frac{(1-\sqrt{1-8t^2})^{|\zeta|}}{(4t)^{|\zeta|}-(1-\sqrt{1-8t^2})^{|\zeta|}}.\end{aligned} \tag{9.13}$$

We denote by ℓ the length of a primitive long geodesic, $\ell \geq 3$ and define $\{m_3, m_4, m_5, \ldots\}$ to be the *length spectrum* of the graph, where m_ℓ is the number of *non-oriented* primitive long geodesics of length ℓ. Next, we shall show that the mean and the variance $\mu(\mathbf{A}, t)$ and $\sigma^2(\mathbf{A}, t)$ can be expressed in terms of the length spectrum of the graph. In particular, we can rewrite (9.13) as

$$\begin{aligned}\mu(\mathbf{A}, t) &= \frac{3\sqrt{1-8t^2}-1}{2(1-9t^2)} + \frac{2}{N}\sum_{\ell=3}^{\infty} \frac{\ell m_\ell}{\sqrt{1-8t^2}} \frac{(1-\sqrt{1-8t^2})^\ell}{(4t)^\ell - (1-\sqrt{1-8t^2})^\ell} \\ &= H(t) + \frac{2}{N}\sum_{\ell=3}^{\infty} m_\ell F_\ell(t), \qquad (9.14)\end{aligned}$$

where

$$H(t) = \frac{3\sqrt{1-8t^2}-1}{2(1-9t^2)}, \quad \text{and} \quad F_\ell(t) = \frac{\ell}{\sqrt{1-8t^2}} \frac{(1-\sqrt{1-8t^2})^\ell}{(4t)^\ell - (1-\sqrt{1-8t^2})^\ell}.$$

By (9.2),

$$\begin{aligned}\sigma^2(\mathbf{A}, t) &= \frac{1}{N}\mathrm{Tr}[(\mathbf{I} - t\mathbf{A})^{-1}(\mathbf{I} - t\mathbf{A})^{-1}] - \mu^2(\mathbf{A}, t) \\ &= \frac{1}{N}\mathrm{Tr}[\frac{d}{dt}(t\mathbf{I} + t^2\mathbf{A} + t^3\mathbf{A}^2 + t^4\mathbf{A}^3 + \cdots)] - \mu^2(\mathbf{A}, t) \\ &= \frac{1}{N}\frac{d}{dt}\left(t\mathrm{Tr}[(\mathbf{I} - t\mathbf{A})^{-1}]\right) - \mu^2(\mathbf{A}, t) \\ &= \mu(\mathbf{A}, t) + t\frac{d}{dt}\mu(\mathbf{A}, t) - \mu^2(\mathbf{A}, t). \qquad (9.15)\end{aligned}$$

By (9.14) and (9.15), we have

$$\begin{aligned}\sigma^2(\mathbf{A}, t) &= H(t) + \frac{2}{N}\sum_{\ell=3}^{\infty} m_\ell F_\ell(t) + t\frac{d}{dt}\{H(t) + \frac{2}{N}\sum_{\ell=3}^{\infty} m_\ell F_\ell(t)\} \\ &\quad - H^2(t) - \{\frac{2}{N}\sum_{\ell=3}^{\infty} m_\ell F_\ell(t)\}^2 - \frac{4}{N}\sum_{\ell=3}^{\infty} m_\ell H(t) F_\ell(t) \\ &= H(t) + tH'(t) - H^2(t) + \frac{2}{N}\sum_{\ell=3}^{\infty} m_\ell F_\ell(t) + \frac{2t}{N}\sum_{\ell=3}^{\infty} m_\ell F'_\ell(t) \\ &\quad - \frac{4}{N^2}\{\sum_{\ell=3}^{\infty} m_\ell F_\ell(t)\}^2 - \frac{4}{N}H(t)\sum_{\ell=3}^{\infty} m_\ell F_\ell(t), \qquad (9.16)\end{aligned}$$

where

$$H'(t) = -3\,\frac{t\left(-5+36\,t^2+3\sqrt{1-8\,t^2}\right)}{\sqrt{1-8\,t^2}\left(-1+9\,t^2\right)^2}.$$

and

$$F'_\ell(t) = \frac{\ell(1-\kappa)^\ell(8(4t)^\ell t^2(\ell+1-\kappa)+8(1-\kappa)^\ell(\kappa-t^2)+(\kappa-1)(4t)^\ell\ell)}{t(1-\kappa)\kappa^3(-(4t)^\ell+(1-\kappa)^\ell)^2},$$

with $\kappa = \sqrt{1-8t^2}$. In particular, for $t = 1/9$, equations (9.14) and (9.16) simplify to, respectively,

$$\mu(\mathbf{A}, 1/9) \approx 1.0395 + \frac{2}{N}\sum_{\ell=3}^{\infty} m_\ell F_\ell(1/9), \text{ and} \tag{9.17}$$

$$\begin{aligned}\sigma^2(\mathbf{A}, 1/9) \approx 0.0433 &+ \frac{2}{N}\sum_{\ell=3}^{\infty} m_\ell F_\ell(1/9) + \frac{2}{9N}\sum_{\ell=3}^{\infty} m_\ell \left[F'_\ell(t)\right]_{t=\frac{1}{9}} \\ &- \frac{4}{N^2}\{\sum_{\ell=3}^{\infty} m_\ell F_\ell(1/9)\}^2 - \frac{4.1580}{N}\sum_{\ell=3}^{\infty} m_\ell F_\ell(1/9),\end{aligned} \tag{9.18}$$

where

$$F_\ell(1/9) = \frac{9}{73}\,\frac{\ell\sqrt{73}\left(1-1/9\sqrt{73}\right)^\ell}{(4/9)^\ell-\left(1-1/9\sqrt{73}\right)^\ell},$$

and

$$\begin{aligned}&\left[F'_\ell(t)\right]_{t=\frac{1}{9}} \\ &= \frac{81}{5329}\,\frac{\ell\rho((584-72\sqrt{73})4^\ell+(72\sqrt{73}-584)\rho+(657\sqrt{73}-5913)4^\ell\ell)}{\left(-9+\sqrt{73}\right)\left(-4^\ell+\rho\right)^2},\end{aligned}$$

with $\rho = (9-\sqrt{73})^\ell$. The results of these derivations, while tedious, can easily be verified using MATLAB or MAPLE.

Rates of Change and Dominant Terms The following analysis of the rates of change, dominant terms, slopes of and distances between filars is similar to arguments presented in Ejov *et al.* [33], in which the self-similar multifilar phenomenon was first discovered.

For various fixed values of $t \in (0, 1/3)$, our experiments show that both function $F_\ell(t)$ and its partial derivative $F'_\ell(t)$ decrease rapidly as ℓ grows. In fact, we observe that $F_\ell\left(\frac{1}{9}\right) \le C_1 10^{-\ell}$ and $\left[F'_\ell(t)\right]_{t=\frac{1}{9}} \le C_2 10^{-\ell}$ for some positive constants C_1 and C_2. It is reasonable to assume that on the other hand, m_ℓ does not grow as fast as $C10^\ell$ for some positive constant C as ℓ

increases, and that $m_{\ell+1}F_{\ell+1} \ll m_\ell F_\ell$ for $\ell \geq 3$.

Consequently, the contribution of the terms of $F_\ell(1/9)$ for $\ell \geq 4$ in (9.18) is insignificant compared to F_3, and thus $\ell = 3$ is the dominant term in the infinite sums in (9.17) and (9.18). Recall our observation that each filar, where each graph has m_3 triangles, is made up of sub-filars. Each of these sub-flars consists of graphs that have exactly m_4 rectangles, with $m_4 = 0$ for the left-most sub-filar and m_4 increases by 1 from one sub-filar to the next sub-filar to the right. Consequently, the lower endpoint of each filar is most likely to contain a graph that has m_3 triangles and zero rectangles. Therefore, from (9 17) and (9.18), with $t = 1/9$, we can approximate the coordinates for the lower end point of each filar with graphs possessing m_3 triangles by

$$\mu(m_3) = 1.0395 + \frac{2}{N} m_3 F_3(1/9), \text{ and} \tag{9.19}$$

$$\begin{aligned}\sigma^2(m_3) = 0.0433 &+ \frac{2}{N} m_3 F_3(1/9) + \frac{2}{9N} m_3 \left[F_3'(t)\right]_{t=\frac{1}{9}} \\ &- \frac{4.1580}{N} m_3 F_3(1/9).\end{aligned} \tag{9.20}$$

Consequently, let $k \in [0, \infty)$, then the line (see Figure 9.3) that goes through the lower end points of filars can be approximated by the parametric line $(x(k), y(k))$, where

$$\begin{aligned}x(k) &= 1.0395 + kF_3(1/9) \\ &\approx 1.0395 + 0.0047k,\end{aligned} \tag{9.21}$$

$$\begin{aligned}y(k) &= 0.0433 + k\{F_3(t) + \frac{1}{9}[F_3'(t)]_{t=\frac{1}{9}} - 2.0790F_3(1/9)\} \\ &\approx 0.0433 + 0.0103k.\end{aligned} \tag{9.22}$$

The slope of the line parametrically described by (9.21) and (9.22) (and represented by the black line in Figure 9.3) is $0.0103/0.0047 = 2.1901$, which is close to the experimental value of 2.04 for cubic graphs of 14 vertices.

The observation that $\ell = 3$ is the dominant term in the infinite sums in (9.17) and (9.18) also explains why in the self-similar structure in Figure 9.2 if two cubic graphs have the same number of triangles, then they belong to the same filar. Similarly, we can explain the membership of sub-filars of various levels. Each filar, where all graphs have m_3 triangles, can be approximated (see Figure 9.4) by a line parametrically defined as follows

$$\begin{aligned}x(s) &= 1.0395 + \frac{2}{N} m_3 F_3(1/9) + sF_4(1/9) \\ &= 1.0395 + \frac{2}{N} 0.0047 m_3 + 0.0007s,\end{aligned} \tag{9.23}$$

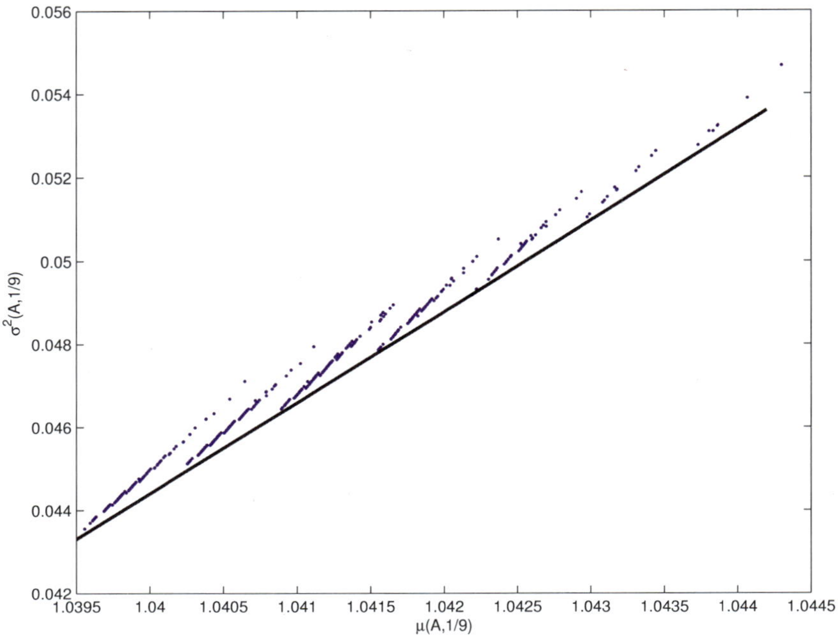

Fig. 9.3: Mean-Variance (trace of resolvent) plot for 14-vertex cubic graphs

$$
\begin{aligned}
y(s) &= 0.0433 + \frac{2}{N} m_3 \{F_3(\tfrac{1}{9}) + \frac{1}{9}\left[F_3'(t)\right]_{t=\frac{1}{9}} - 2.0790 F_3(1/9)\} \\
&\quad + s\{F_4(\tfrac{1}{9}) + \frac{1}{9}\left[F_4'(t)\right]_{t=\frac{1}{9}} - 2.0790 F_4(1/9)\} \\
&= 0.0433 + \frac{2}{N} 0.0103 m_3 + 0.0023 s.
\end{aligned} \tag{9.24}
$$

The slope of each filar is approximately

$$
\frac{F_4(\tfrac{1}{9}) + \frac{1}{9}\left[F_4'(t)\right]_{t=\frac{1}{9}} - 2.0790 F_4(1/9)}{F_4(1/9)} \approx 3.2448,
$$

which is independent of the graph order N.

Consider two consecutive filars, consisting of graphs containing exactly $m_3^{(1)}$ and $m_3^{(2)}$ triangles respectively. Then, the line approximating the first filar is parametrically defined by

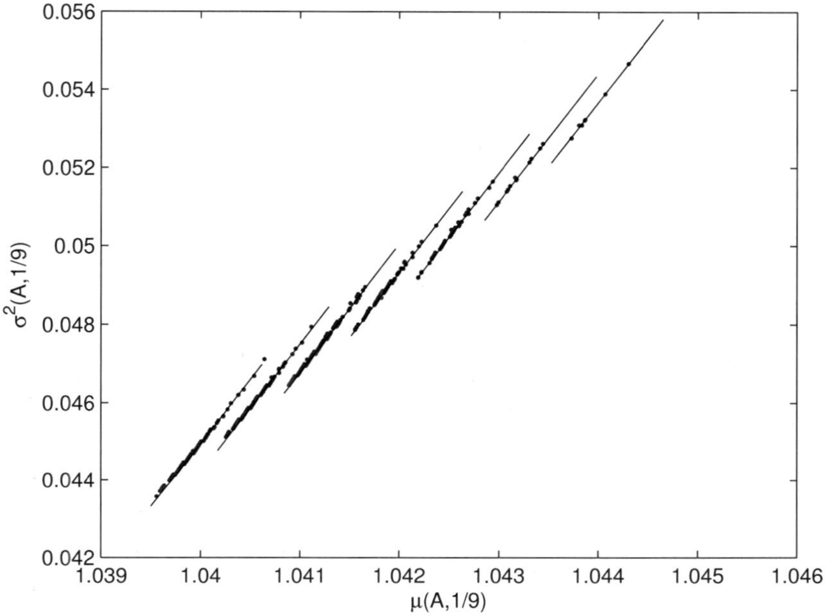

Fig. 9.4: Mean-Variance (trace of resolvent) plot for 14-vertex cubic graphs

$$\bar{x}(s_1) = 1.0395 + \frac{2}{N}0.0047m_3^{(1)} + 0.0007s_1,$$
$$\bar{y}(s_1) = 0.0433 + \frac{2}{N}0.0103m_3^{(1)} + 0.0023s_1,$$

and the line approximating the second filar is parametrically defined by

$$\hat{x}(s_2) = 1.0395 + \frac{2}{N}0.0047m_3^{(2)} + 0.0007s_2,$$
$$\hat{y}(s_2) = 0.0433 + \frac{2}{N}0.0103m_3^{(2)} + 0.0023s_2,$$

In order to find out the horizontal distance between two filars approximated by the parametric lines $(\bar{x}(s_1), \bar{y}(s_2))$ and $(\hat{x}(s_1), \hat{y}(s_2))$, firstly, we need to find out s_1 and s_2 such that $\bar{y}(s_1) = \hat{y}(s_2)$

$$0.0433 + \frac{2}{N}0.0103m_3^{(1)} + 0.0023s_1 = 0.0433 + \frac{2}{N}0.0103m_3^{(2)} + 0.0023s_2$$
$$s_1 = s_2 + 0.0206\frac{m_3^{(2)} - m_3^{(1)}}{N}.$$

Then the horizontal distance between the two aforementioned filars is

$$\begin{aligned}
&\hat{x}(s_2) - \bar{x}(s_2 + 0.0206\frac{m_3^{(2)} - m_3^{(1)}}{N}) \\
&= \frac{2}{N}0.0047m_3^{(2)} + 0.0007s_2 - \frac{2}{N}0.0047m_3^{(1)} - 0.0007(s_2 + 0.0206\frac{m_3^{(2)} - m_3^{(1)}}{N}) \\
&= 0.0094\frac{m_3^{(2)} - m_3^{(1)}}{N}.
\end{aligned}$$

Hence, the horizontal distance between two consecutive filars decreases as the graph order N increases, so the filars are closer to each other as the graphs get larger.

9.4 Self-similarity and Hamiltonicity

Next, we reconstruct the variance-versus-mean plot in Figure 9.2, but distinguish between Hamiltonian and non-Hamiltonian graphs. Each pair of coordinates $(\mu_{\mathbf{A}}(1/9), \sigma^2_{\mathbf{A}}(1/9))$ of a 14-vertex cubic graph is a dot on this reconstructed plot (see Figure 9.5) if the graph is Hamiltonian, and a cross if the graph is non-Hamiltonian. The discussion of this section is based on empirical evidence only.

Two graphs are *cospectral* if they share the same spectrum, and are *non-cospectral* otherwise. There are no cospectral cubic graphs with fewer than 14 vertices, and there are at least three pairs of cospectral cubic graphs on 14 vertices [56]. Therefore, there are three pairs of graphs of which the coordinates in the above plot are the same. However, as these three pairs are all Hamiltonian graphs, their cospectral property does not affect our observations on the self-similar multifilar structure and Hamiltonicity.

Glancing at the plot in Figure 9.5, it is easy to think that while the majority of non-Hamiltonian graphs are located in the top and bottom parts of filars, some of them are mixed among dots representing Hamiltonian graphs. However, if we zoom in on the innermost[1] sub-filars, we will find that all non-Hamiltonian graphs are strictly at the top and the bottom of these sub-filars, and there is no mixing between Hamiltonian and non-Hamiltonian graphs. While it is not practical to show all plots which zoom in on the innermost sub-filars, we present here one of these zooming-in plots (see Figure 9.6) to illustrate our observation. All crosses that can be seen clearly in this plot are either at the top or the bottom of their sub-filars.

---

[1] When we can no longer zoom in on a filar to obtain a similar structure made up of smaller filars, we are at the innermost sub-filars.

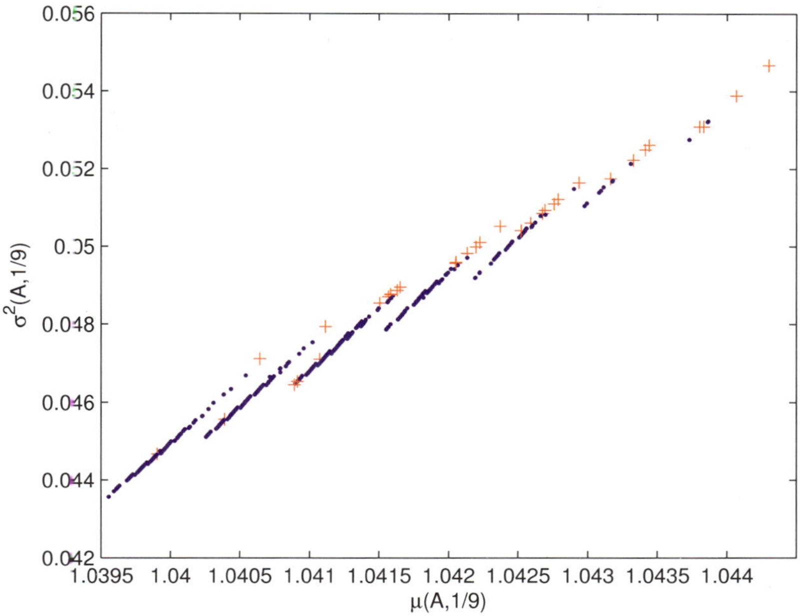

Fig. 9.5: Mean-Variance plot for 14-vertex cubic graphs and Hamiltonicity

We have not yet been able to explain why the absence of Hamiltonian cycles makes non-Hamiltonian graphs gather around the top and the bottom of their innermost sub-filars. However, we have experimentally found the answer as to which non-Hamiltonian graphs are at the higher end while the rest are at the lower.

Briefly, non-Hamiltonian graphs that are located at the top of their sub-filars are *bridge graphs*. A bridge graph is a graph that contains at least one *bridge*, that is, an edge the removal of which disconnects the graph. Sainte Lague [72, Section 57] states a very natural theorem: all bridge graphs are non-Hamiltonian. These graphs can be identified in polynomial time; hence, we refer to them as *easy non-Hamiltonian graphs*. A non-Hamiltonian graph that is not an easy non-Hamiltonian graph is a *hard non-Hamiltonian graph*. The latter group are found to be at the bottom of their sub-filars. These hard non-Hamiltonian graphs constitute the underlying difficulty of the NP-hard complexity to the Hamiltonian cycle problem. For example, Figure 9.7 shows the plot of coordinates $(\mu_{\mathbf{A}}(1/9), \sigma^2_{\mathbf{A}}(1/9))$ for 10-vertex (connected) cubic graphs, of which there are 19, including 17 Hamiltonian ones.

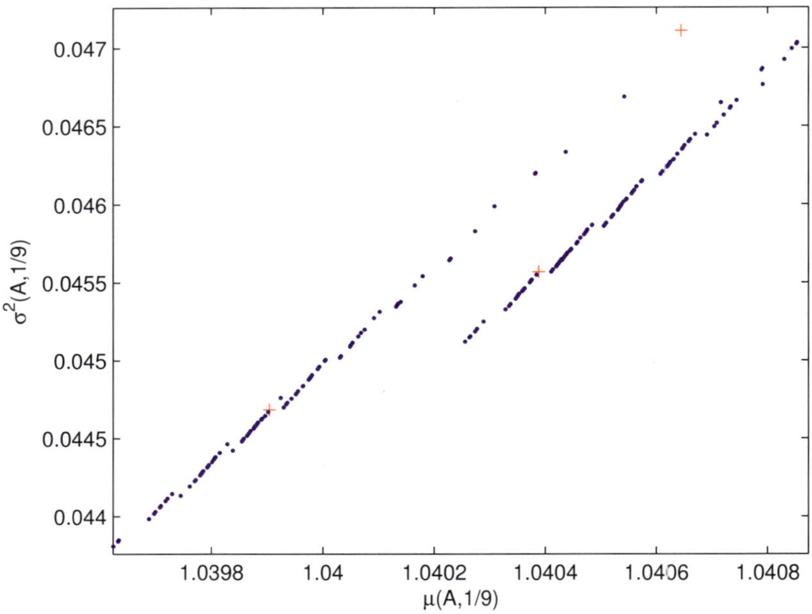

Fig. 9.6: Mean-Variance plot—zooming in

One non-Hamiltonian graph is a bridge graph (see Figure 9.8), represented by the cross at the top right of the plot. The other non-Hamiltonian graph is the well-known Petersen graph (see Figure 9.9), which is not a bridge graph and is represented by the cross at the bottom left of the same plot.

It would be interesting to obtain a theoretical justification of this observation.

From numerical experiments using GENREG and *nauty* algorithms (see Meringer [79] and McKay [78], respectively) on cubic graphs of various orders up to $N = 24$, we have observed that bridge graphs make up the majority of non-Hamiltonian graphs. Moreover, as the graph order N increases, so does the ratio of cubic bridge graphs over all cubic non-Hamiltonian graphs of the same order. This can be seen from Table 9.1.

For cubic graphs of order 40 and 50, we consider a 1000000-graph sample for each order. The observed ratios of cubic bridge graphs to cubic non-Hamiltonian graphs in Table 9.2 are even closer to 1. This naturally gives rise to a conjecture on the prevalence of cubic bridge graphs.

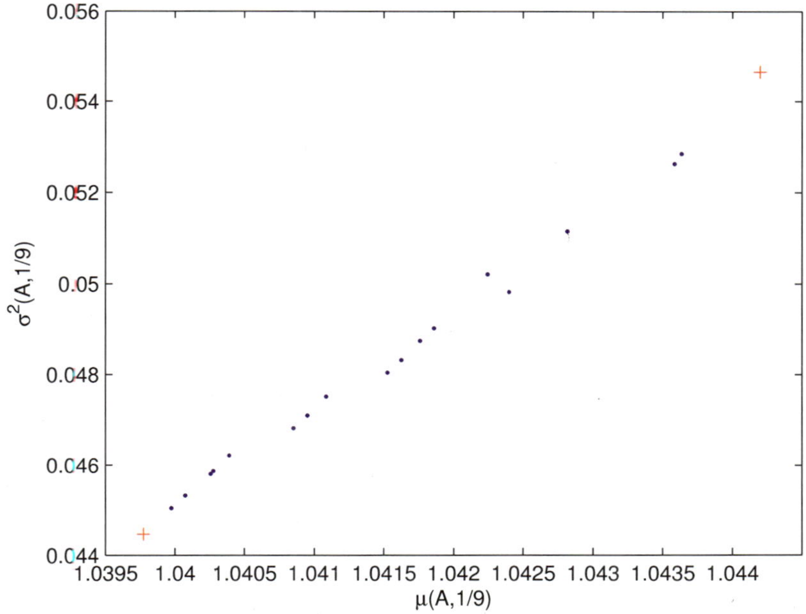

Fig. 9.7: Mean-Variance plot for 10-vertex cubic graphs and Hamiltonicity

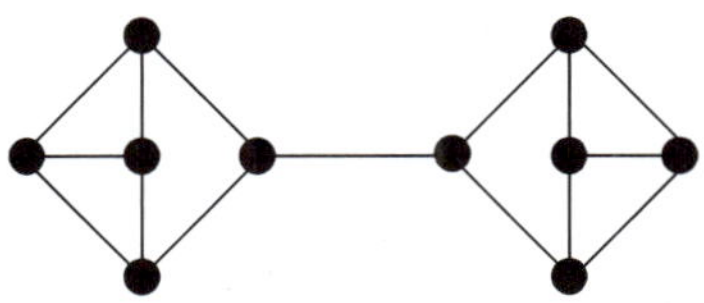

Fig. 9.8: The only cubic bridge graph of order 10

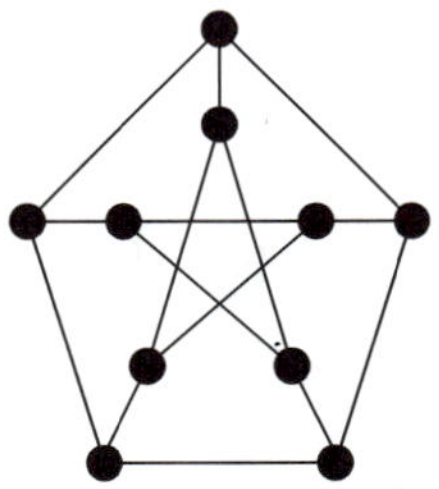

Fig. 9.9: Petersen graph

Conjecture 9.1. [48] *Consider cubic graphs of order N.*

$$\lim_{N\to\infty} \frac{\#\text{cubic bridge graphs}}{\#\text{cubic non-Hamiltonian graphs}}$$
$$= \lim_{N\to\infty} \frac{\#\text{easy non-Hamiltonian graphs}}{\#[\text{easy non-Hamiltonian graphs} + \text{hard non-Hamiltonian graphs}]} = 1.$$

In summary, the study of the self-similar, multifilar, structure of cubic graphs revealed not only that bridge graphs cluster in the upper ends of individual

Table 9.1: Ratio of cubic bridge graphs over cubic non-Hamiltonian graphs

N	Cubic	Cubic Non-H	Cubic Bridge	Bridge/Non-H
10	19	2	1	0.5000
12	85	5	4	0.8000
14	509	35	29	0.8286
16	4060	219	186	0.8493
18	41301	1666	1435	0.8613
20	510489	14498	12671	0.8740
22	7319447	148790	131820	0.8859
24	117940535	1768732	1590900	0.8995

Table 9.2: Ratio of cubic bridge graphs over cubic non-Hamiltonian graphs

N	Cubic	Cubic Non-H	Cubic Bridge	Bridge/Non-H
40	1000000	912	855	0.9375
50	1000000	549	530	0.9650

filars but also that, as N increases, they seem to vastly outnumber remaining non-Hamiltonian graphs. Conjecture 9.1 is based on the belief that as N grows, the population of non-Hamiltonian cubic graphs will be dominated by bridge graphs as it is rather easy to construct bridge graphs of a given order from many pairs of cubic graphs of smaller orders. For instance, if N is given we can take any two connected cubic graphs G_1 and G_2 of orders N_1 and N_2, respectively, such that $N = N_1 + N_2 - 2$, and select an edge e_1 from G_1 and e_2 from G_2. Then, by inserting a new vertex in the middle of each of those two edges and joining these two vertices by a new edge, we create a bridge graph of order N. Thus, it is natural to expect that bridge graphs may have a competitive advantage in spreading across the population of non-Hamiltonian cubic graphs. Unfortunately, it is not clear how to turn this kind of intuitive reasoning into a rigorous argument. However, in Chapter 10 we present some graph enumeration results that could, perhaps, be useful in resolving this conjecture.

Chapter 10
Graph Enumeration

10.1 Introduction

Graph enumeration is a study in graph theory that deals with counting non-isomorphic graphs with a particular property. Harary and Palmer [60] provide an excellent introduction to the topic of graph enumeration. On counting labelled cubic graphs, there has been a series of results, most notably Read [88], Read [89], Wormald [104], and Wormald [105], which collectively present various approaches for counting labelled cubic graphs, and labelled cubic graphs with a given connectivity. In comparison to labelled cubic graphs, the numeration of unlabelled cubic graphs is a significantly more challenging problem [105]. Robinson [90] presents a method to count unlabelled cubic graphs.

In this chapter, we present the starting point of a possible approach to resolving Conjecture 9.1 on the prevalence of cubic bridge graphs. The idea is, if we can find a method, possibly recursive, to count all cubic bridge graphs of a given order, and use the same method to count the number of cubic bridge graphs of the same order, then we can evaluate the ratio of cubic bridge graphs to cubic graphs. Then, we can compare this ratio to the ratio determined in Robinson and Wormald [91], which also states the striking result that almost all cubic graphs are Hamiltonian.

Here, we derive a new recursive formula for the cardinality of unlabelled cubic graphs with bridges, weighted by the number of orbits of bridges in each graph. In the process of deriving the formula, we introduce a new graph property: *subdivision-equivalent edges*. Two edges in a graph are *subdivision-equivalent* if the two graphs G and H obtained by subdividing each edge respectively are isomorphic. In a sense, subdivision-equivalent edges are the opposite of *pseudo-similar edges*. Two edges in a graph are *pseudo-similar* if the two graphs G and H obtained by removing each edge respectively are isomorphic, but there is no automorphism of the original graph that maps one edge to another. Pseudo-similar edges have received a lot of attention due to their relationship to the well-known Reconstruction Conjecture in graph theory [75], which states that graphs are uniquely determined by their set of subgraphs.

10.2 Subdivision-equivalent Edges

Wormald [105] presents formulae to enumerate labelled cubic graphs with a given connectivity. In particular, the author determines recurrence relations to count connected cubic graphs, 2-connected cubic graphs and 3-connected cubic graphs. Recall that all cubic bridge graphs are 1-connected, but not 2- or 3-connected. Consequently, subtracting the number of labelled 2-connected cubic graphs from the total number of labelled cubic graphs of the same order gives the number of labelled cubic bridge graphs. This, together with the fact that the expected number of non-trivial automorphisms of a labelled cubic graph of order N approaches zero as N tends to infinity [106], gives us an asymptotic number of unlabelled bridge graphs. However, subtracting one recurrence relation from another recurrence relation, in this case, results in a formula that is not easy to interpret or manipulate. For the sake of completeness, we include here the two aforementioned recurrence relations.

Proposition 10.1. [105] *For $N = 2n$, the following two statements hold.*
(i) The number of labelled cubic graphs of order N is

$$\frac{(2n)!}{3n2^n}\hat{r}_n,$$

where $\hat{s}_0 = \hat{s}_1 = 0, \hat{s}_2 = 1$, and

$$\begin{aligned}\hat{s}_n &= 3n\hat{s}_{n-1} + 4\hat{s}_{n-2} + 2\hat{s}_{n-3} \\ &\quad + \sum_{i=2}^{n-3} \hat{s}_i \left(\hat{s}_{n-1-i} - 2\hat{s}_{n-2-i} - 2\hat{s}_{n-3-i}\right), \quad \textit{for } n \geq 3, \\ \hat{r}_n &= \hat{s}_n - 2\hat{s}_{n-1} - 2\hat{s}_{n-2}, \quad \textit{for } n \geq 2.\end{aligned}$$

(ii) The number of labelled 2-connected cubic graphs of order N is

$$\frac{(2n)!}{3n2^n}\widetilde{r}_n,$$

where $\widetilde{s}_1 = 0, \widetilde{s}_2 = 1$, and

$$\widetilde{s}_n = 3n\widetilde{s}_{n-1} + 2\widetilde{s}_{n-2} + (3n-1)\sum_{i=2}^{n-3} \widetilde{s}_1\widetilde{s}_{n-1-i}, \quad \textit{for } n \geq 3,$$

$$\widetilde{r}_n = \widetilde{s}_n - 2\widetilde{s}_{n-1}, \qquad for\ n \geq 2.$$

Automorphisms, Isomorphisms, Orbits and Similarities Consider two graphs G and H. Let $V(G)$ and $V(H)$ be the sets of vertices, and $E(G)$ and $E(H)$ be the sets of edges in G and H, respectively. Recall that the graphs G and H are *isomorphic* if there exists a bijection $f : V(G) \mapsto V(H)$ such that for every edge $(u, v) \in E(G)$, the edge $(f(u), f(v)) \in E(H)$, and they are said to be *non-isomorphic* otherwise. An *automorphism* of a graph G is an *isomorphism* of G with itself [59]. Two edges e_1 and e_2 in G are *similar* if there is an automorphism that maps e_1 to e_2. Similar edges are said to be in the same *orbit*. An orbit of edges in G is a set of edges in G such that every edge in the set is similar to any other edge in the set.

Example 10.1 *The following two graphs G and H are isomorphic:*

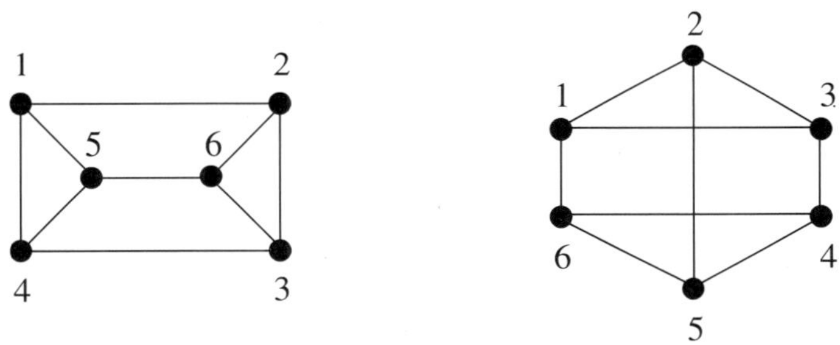

Fig. 10.1: G and H are isomorphic

This is because there exists at least one bijection f*:* $V(G) \mapsto V(H)$*, with* $f(1) = 6, f(2) = 1, f(3) = 3, f(4) = 4, f(5) = 4,$ *and* $f(6) = 2$*, such that for every* $(u, v) \in E(G)$*, the edge* $(f(u), f(v)) \in E(H)$*. An automorphism of* G *is* $g : V(G) \mapsto V(G)$*, with* $g(1) = 3, g(2) = 4, g(3) = 1, g(4) = 2, g(5) = 6,$ *and* $g(6) = 5$*. Under this particular automorphism, the edge* $(1, 2)$ *is mapped into* $(3, 4)$*, so they are similar. These two edges are also in the same orbit. Note that there are other edges in this orbit, and there are other orbits of edges in this graph.*

Pseudo-Similarity and Removal-Similarity. Let $G - e_2$ be the original graph G with the edge e_2 removed. Then $V(G) = V(G - e_2)$ and $E(G) = E(G - e_2)\backslash\{e_2\}$. If e_1 and e_2 are similar, then $G - e_1$ and $G - e_2$ are isomorphic graphs. Two edges e_1 and e_2 are *pseudosimilar* if $G - e_1$ and $G - e_2$ are isomorphic graphs but e_1 and e_2 are not similar in G. *Edge-similarity* and *edge-pseudosimilarity* are collectively known as *edge-removal-similarity*. The parallel concepts of *vertex-similarity*, *vertex-pseudo-similarity* and *vertex-removal-similarity* are defined analogously. As we are only concerned about edge-similarity, edge-pseudosimilarity, and edge-removal-similarity, for convenience, we will drop the prefix *edge-* whenever confusion cannot arise.

We present here a recursive formula, (10.1), to determine the number of cubic bridge graphs of a given order N, weighted by the number of orbits of bridges in these graphs. This means that, when counting cubic bridge graphs of any given order, we count the number of cubic bridge graphs that have one orbit of bridges once, two orbits of bridges twice, and so on.

Remark 10.1. For example, consider three cubic bridge graphs G_1, G_2 and G_3, where G_1 has two bridges, G_2 has three bridges, and G_3 has one bridge. Furthermore, assume that two bridges in G_1 are not in the same orbit (hence G_1 has two orbits of bridges), that all three bridges in G_2 are in the same orbit (hence G_2 has one orbit of bridges). In our method, we count G_1 twice, G_2 once and G_3 once. Hence, the total number of cubic bridge graphs of a given order N, weighted by the number of orbits of bridges in these graphs, is an overestimate of the total number of N-vertex cubic bridge graphs. An open problem is to estimate the difference between these two counts, which we conjecture to be approaching zero as N tends to infinity.

In the process of deriving the aforementioned formula (10.1), we introduce the useful notion of *subdivision-equivalent* edges. Consider an edge $e = (s,t) \in E(G)$. The edge e is *subdivided* if we add a new vertex w to $V(G)$ and replace the edge e with two edges (s,w) and (w,t).

Definition 10.2. *Given a graph G, let G_1 and G_2 be two resulting graphs obtained by subdividing edges e_1 and e_2 of G, respectively. Then, the two edges e_1 and e_2 are* subdivision-equivalent *if and only if G_1 and G_2 are isomorphic.*

Let $\mathsf{Eqe}(G)$ denote the number of sets of subdivision-equivalent edges in G, and let $\mathsf{Eqe}(\mathcal{R})$ denote $\sum_i \mathsf{Eqe}(R_i)$, if $\mathcal{R}$ is a set of graphs R_i.

Example 10.3 *Consider the labelled envelope graph G.*

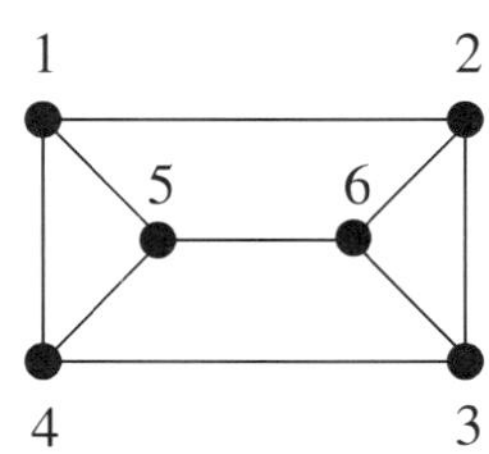

Fig. 10.2: The labelled envelope graph G

Subdividing $(1,2)$ and $(3,4)$ gives us the two graphs G_1 and G_2 in Figs 10.3 and 10.4, respectively. In G, the edges $(1,2)$ and $(3,4)$ are subdivision-equivalent, as the resulting graphs G_1 and G_2 obtained after subdividing these edges respectively are isomorphic. There are other pairs of subdivision-equivalent edges in this graph.

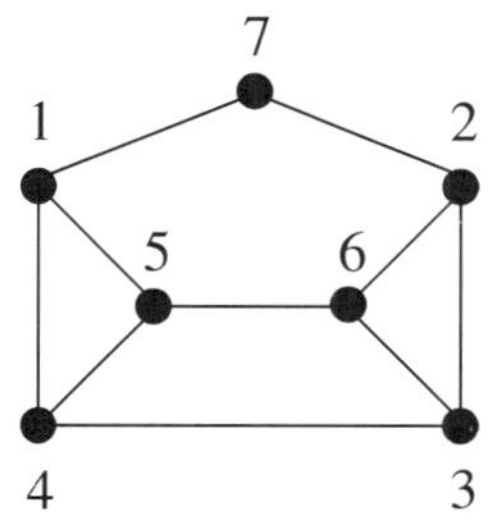

Fig. 10.3: G_1

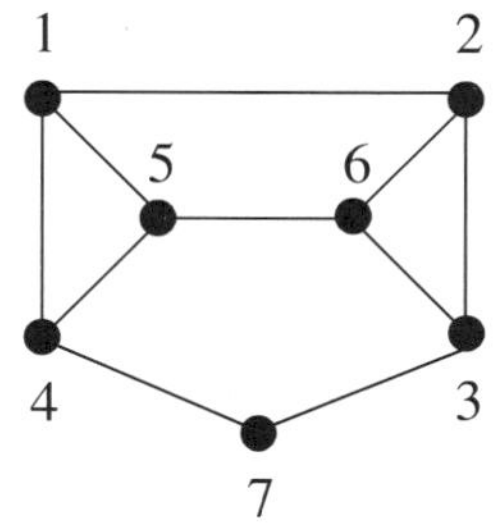

Fig. 10.4: G_2

One of our, still open, problems concerns the question of whether, for regular graphs, two edges are subdivision-equivalent if and only if they are removal-similar.

However, in a general graph, subdivision-equivalent edges are not necessarily removal-similar and vice versa. The following counterexample was provided by Brendan McKay, through private communication. We consider two graphs: in the first one, a pair of edges are removal-similar but not subdivision-equivalent, and in the second, a pair of edges are subdivision-equivalent but not removal-similar.

Example 10.4 *Edges a and c in G are removal-similar but not subdivision-equivalent:*

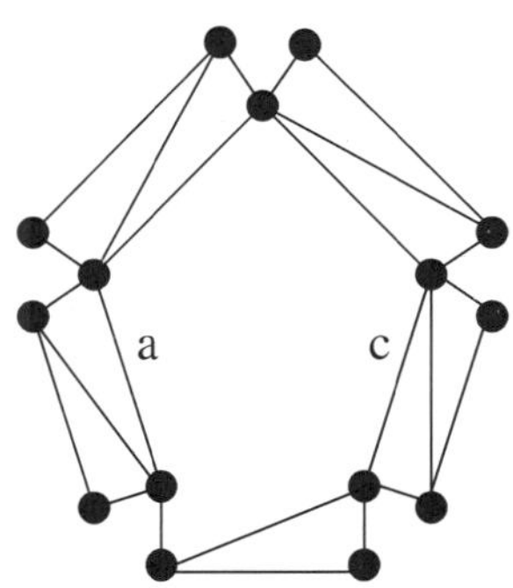

Fig. 10.5: Graph G

The resulting graphs after removing a and c, respectively, are isomorphic (see Figure 10.6), but the resulting graphs after subdividing a and c, respectively, are non-isomorphic (see Figure 10.7).

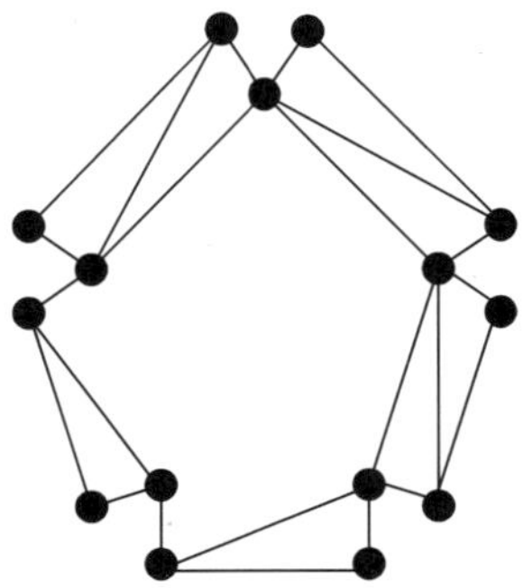

Fig. 10.6: The resulting graphs are isomorphic

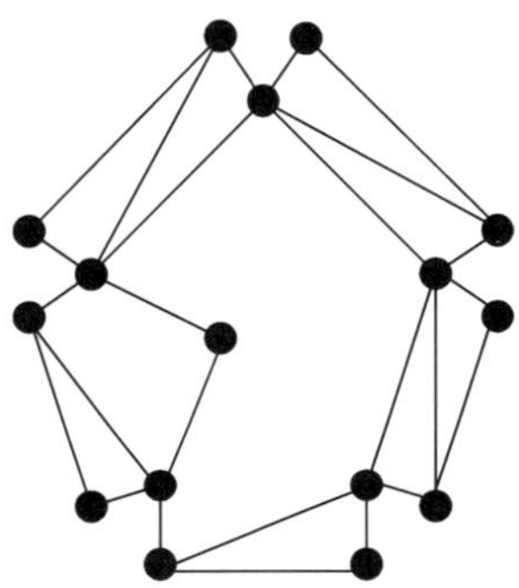

Fig. 10.7: The resulting graphs are non-isomorphic

In the graph Π below, the edges a and c are subdivision-equivalent but not removal-similar:

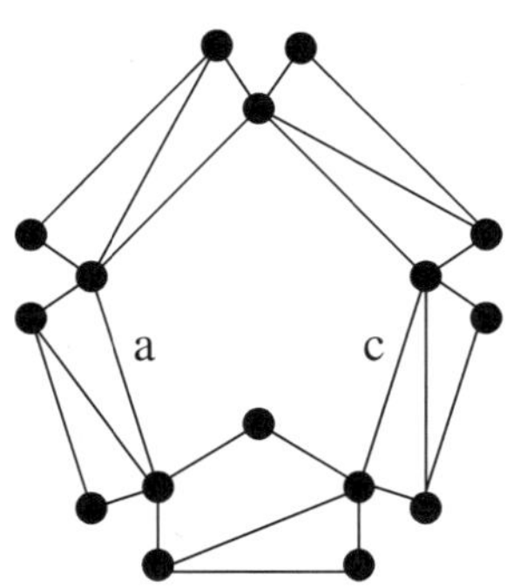

Fig. 10.8: Graph Π

The resulting graphs and after subdividing a and c, respectively, are isomorphic:

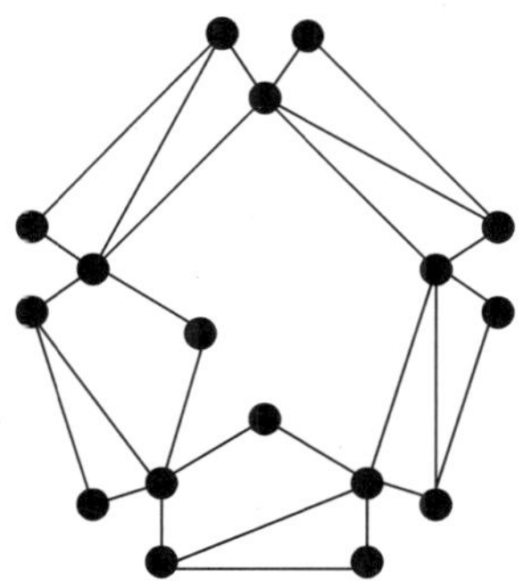

Fig. 10.9: The resulting graphs are isomorphic

The resulting graphs after removing a and c, respectively, are non-isomorphic:

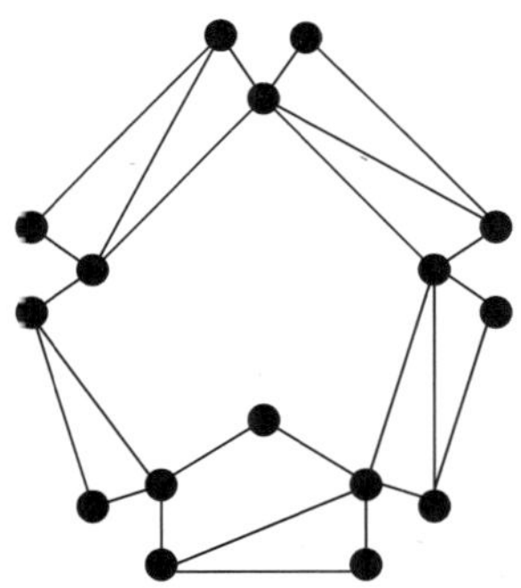

Fig. 10.10: The resulting graphs are non-isomorphic

10.3 Enumerating Cubic Bridge Graphs

We denote by C_N and B_N respectively the sets of cubic graphs and cubic bridge graphs of order N. Our recursive formula for the total number of cubic bridge graphs of order N, weighted by the number of orbits of bridges in each graph, is dependent on $\mathsf{Eqe}(C_{M_i})$ and the number of cubic bridge graphs of various orders $M < N$.

Consider a cubic bridge graph G of order N with a bridge (u, v), displayed in Figure 10.11. Let us denote two vertices that are connected to u (beside v) by u_1 and u_2, and two vertices that are connected to v (beside u) by v_1 and v_2.

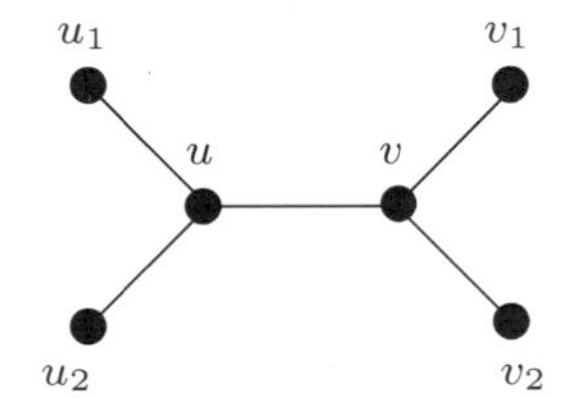

Fig. 10.11: A bridge (u, v)

When a vertex t is removed from a graph, it is assumed that all edges with t as one of its ends are also removed from the graph. Let G_1 and G_2 be the two resulting, disjoint, components if we remove two vertices u and v, and assume that $u_1 \in V(G_1)$ while $v_1 \in V(G_2)$. As (u, v) is a bridge in G, we have $V(G_1 \cup G_2) = V(G) \backslash \{u, v\}$, $u_2 \in V(G_1)$, $v_2 \in V(G_2)$ and $|V(G_1)| + |(V(G_2)| + 2 = N$.

Example 10.5 *Recall the only cubic bridge graph of order 10:*

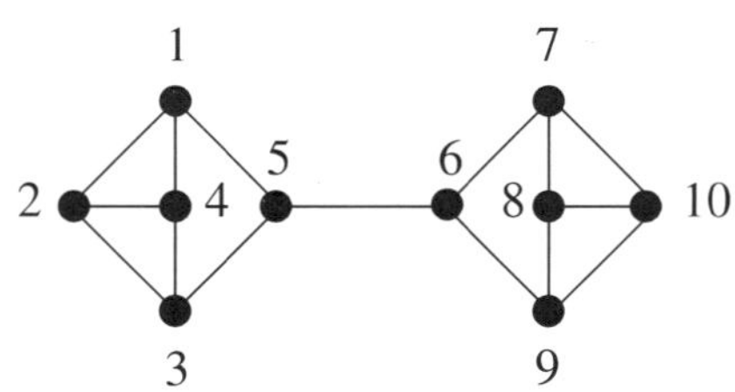

Fig. 10.12: The only cubic bridge graph of order 10

In this graph, the edge $(5, 6)$ is a bridge. Removing this bridge disconnects the original graph, resulting in two separate components G_1 and G_2, in Figures 10.13 and 10.13, respectively.

In any cubic graph of order N, the number of edges is $3N/2$. Therefore, N has to be even, and $|V(G_1)|$ and $|V(G_2)|$ have to be both odd or both even.

Lemma 10.1. *$|V(G_1)|$ and $|V(G_2)|$ are both even.*

Proof. Suppose $|V(G_1)|$ and $|V(G_2)|$ are both odd. Let $|V(G_1)| = 2k+1, k \geq 2$. In $V(G_1)$, there are $2k - 1$ vertices with degrees of three and two vertices,

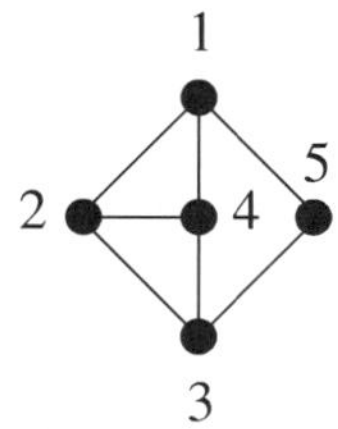

Fig. 10.13: G_1

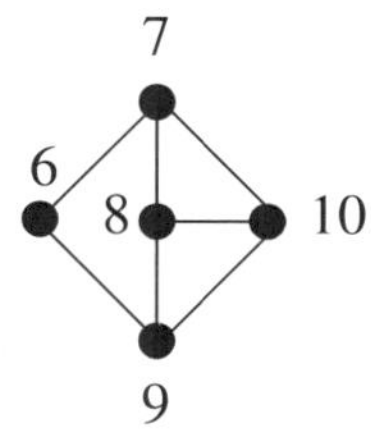

Fig. 10.14: G_2

u_1 and u_2, with degrees of two. The number of edges in G_1 is $(3(2k-1)+4)/2$, which implies that $2k-1$ is an even number, resulting in a contradiction. □

For $k, q \geq 2$, let $|V(G_1)| = 2k$ and $|V(G_1)| = 2q$, where $2k + 2q + 2 = N$. Note that it is not possible to construct G_1 (or G_2) if $k = 1$ (or $q = 1$). Consequently, smallest possible cubic bridge graphs are of order 10. Recall that, in this case, a bridge graph is one of two cubic non-Hamiltonian graphs of order 10, the other one being the famous Petersen graph.

Theorem 10.1. *The number of cubic bridge graphs of order N, weighted by the number of orbits of bridges in each graph, is*

$$\sum_i f(k_i, q_i) + |B_{N-4}| + |B_{N-2}|, \tag{10.1}$$

where (k_i, q_i) are all possible unordered pairs of integers $k_i, q_i \geq 2, 2k_i + 2q_i + 2 = N$, and

$$f(k_i, q_i) = \begin{cases} \mathsf{Eqe}(C_{2k_i})\mathsf{Eqe}(C_{2q_i}) \text{ if } k_i \neq q_i, \\ \frac{1}{2}\mathsf{Eqe}(C_{2k_i})\,[\mathsf{Eqe}(C_{2k_i}) + 1] \text{ if } k_i = q_i. \end{cases} \tag{10.2}$$

Proof. Consider a cubic bridge graph G of order N with a bridge (u, v) as described in Figure 10.11. Again, let G_1 and G_2 be the two resulting, disjoint, components if we remove two vertices u and v and without loss of generality, assume that $u_1 \in V(G_1)$ while $v_1 \in V(G_2)$.

There are now three cases to consider:

1. both edges (u_1, u_2) and (v_1, v_2) are not in $E(G)$,
2. only one of them is $E(G)$, and
3. both edges are in $E(G)$.

For $i = 1, 2, 3$, let B_N^i denote the set of cubic bridge graphs that have an orbit of bridges in each Case i. We count the number of cubic bridge graphs of order N that have an orbit of bridges of Cases 1, 2, and 3, respectively. Naturally, a cubic bridge graph might have two or more orbits of bridges, each of which might be in a different or the same case. In an abuse of terminology, we will simply refer a cubic bridge graph that has an orbit of bridges of Case i as *a cubic bridge graph of Case i*. Adding these three numbers together gives us the total number of cubic bridge graphs of order N where each graph with k orbits of bridges is counted k times.

For example, consider a cubic bridge graph H_1 with three bridges, none of which is in the same orbit of another. Therefore, H_1 has three orbits of bridges. Assume that each of these bridges is in a different case, then H_1 is a cubic bridge graph of Case 1, of Case 2 and also of Case 3. In our counting, H_1 is counted three times, once in each case. Consider a different cubic bridge graph H_2 with three bridges e_1, e_2 and e_3, where e_1 and e_2 are in the same orbit, and e_3 is in a different orbit, but both orbits of bridges are in Case 1. Then H_2 is counted twice in Case 1.

Case 1. Here, $(u_1, u_2), (v_1, v_2) \notin E(G)$.

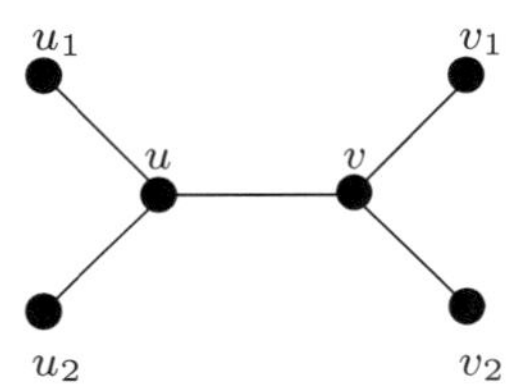

Fig. 10.15: Neighbour around the bridge (u, v)—Case 1

In G_1 and G_2, inserting two new edges (u_1, u_2) and (v_1, v_2) results in two cubic graphs $\bar{G}_1$ and $\bar{G}_2$ of orders $2k$ and $2q$, respectively. Therefore, if we choose any two cubic graphs of orders $2k$ and $2q$, subdivide an edge in each graph (consequently introducing two new vertices), and connect these two vertices (thereby creating a bridge), we obtain a cubic bridge graph of order $2k + 2q + 2 = N$. For each G, the number of non-isomorphic graphs obtained by subdividing an edge is precisely the number $\mathsf{Eqe}(G)$ of sets of equivalent edges in G.

Therefore,

$$|B_N^1| = \sum_{i=1}^{t} f(k_i, q_i), \tag{10.3}$$

where t is the number of possible unordered pairs of integers (k_i, q_i) such that $2k_i + 2q_i + 2 = N$; $k_i, q_i \geq 2$, and

$$f(k_i, q_i) = \begin{cases} \mathsf{Eqe}(C_{2k_i})\mathsf{Eqe}(C_{2q_i}), & \text{if } k_i \neq q_i, \\ \frac{1}{2}\mathsf{Eqe}(C_{2k_i})\left[\mathsf{Eqe}(C_{2k_i}) + 1\right], & \text{if } k_i = q_i. \end{cases} \tag{10.4}$$

Note that the difference in the formulae of the case for $k_i \neq q_i$ and the case for $k_i = q_i$ is due to the fact that, for the former, we are counting pairs of objects from two different sets, whereas for the latter, we are counting pairs of objects from the same set.

Case 2. Here, $(u_1, u_2) \in E(G), (v_1, v_2) \notin E(G)$. Let $u_3, u_4 \notin \{u, u_2\}$ be the third vertices to which u_1 and u_2 are adjacent, respectively. Now there are two subcases: $u_3 = u_4$ (Figure 10.16) and $u_3 \neq u_4$ (Figure 10.17).

Case 2a. Here, $u_3 = u_4$. Let $u_5 \notin \{u_1, u_2\}$ be the third vertex to which u_3 is adjacent.

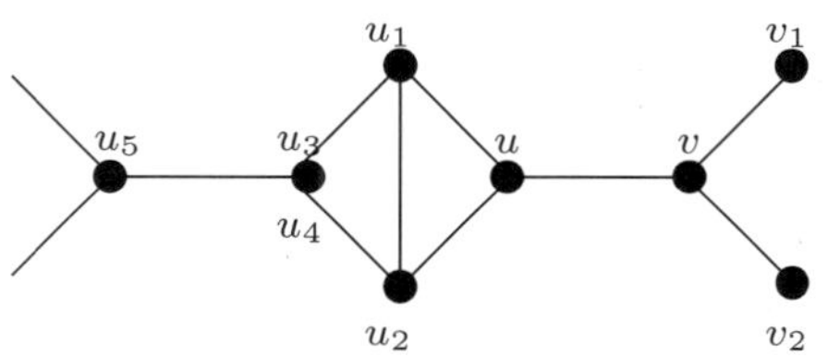

Fig. 10.16: Neighbour around the bridge (u, v) in G—Case 2a

Since $V(G_1) \cap V(G_2) = \emptyset$, (u_3, u_5) is also a bridge. Contracting G by removing the set of "diamond" vertices $\{u, u_1, u_2, u_3\}$ and connecting v to u_5 results in a bridge graph of order $N - 4$ which belongs to either Case 1 or Case 2, since $(v_1, v_2) \notin E(G)$. Therefore, if we choose any cubic bridge graph of order $N - 4$ that belongs to either Case 1 or Case 2, insert a diamond set of four vertices and their associated edges like $\{u, u_1, u_2, u_3\}$ depicted in Figure 10.16 into the bridge, we obtain a cubic bridge graph of order N.

Therefore,

$$|B_N^{2a}| = |B_{N-4}^1| + |B_{N-4}^2|. \tag{10.5}$$

Since the smallest possible cubic bridge graphs are of order 10, the smallest possible cubic bridge graphs in Case 2a are of order 14.

Case 2b. Here, $u_3 \neq u_4$ (see Figure 10.17). Contracting G by removing the set of vertices $\{u_1, u_2\}$ and connecting u to u_3 and u_4 results in a bridge graph of order $N-2$ which belongs to either Case 1 or Case 2, since $(v_1, v_2) \notin E(G)$. Therefore, if we choose any cubic bridge graph of order $N - 2$ that belongs to either Case 1 or Case 2, insert two vertices and their associated edges like $(u_1, u_2), (u_1, u)$ and (u_2, u) depicted in Figure 10.17 into one side of the bridge (the side that has a connecting edge between two vertices), we obtain a cubic bridge graph of order N.

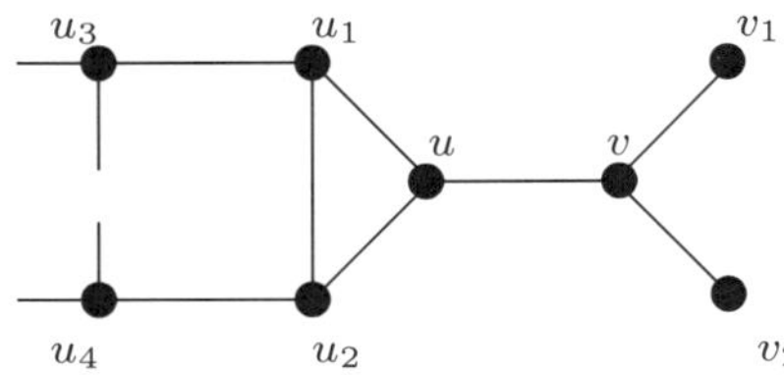

Fig. 10.17: Neighbour around the bridge (u, v) in G—Case 2b

Therefore,

$$|B_N^{2b}| = |B_{N-2}^1| + |B_{N-2}^2|. \tag{10.6}$$

Since the smallest possible cubic bridge graphs are of order 10, the smallest possible cubic bridge graphs in Case 2b are of order 12.

Case 3. Here, $(u_1, u_2), (v_1, v_2) \in E(G)$. Let $u_3, u_4 \notin \{u, u_2\}$ be the "third vertices" to which u_1 and u_2 are adjacent, respectively. Now there are two subcases: $u_3 = u_4$ and $u_3 \neq u_4$.

Case 3a. Here, $u_3 = u_4$. This case can be analysed in a manner similar to Case 2a. Let $u_5 \notin \{u_1, u_2\}$ be the "third vertex" that u_3 is adjacent to.

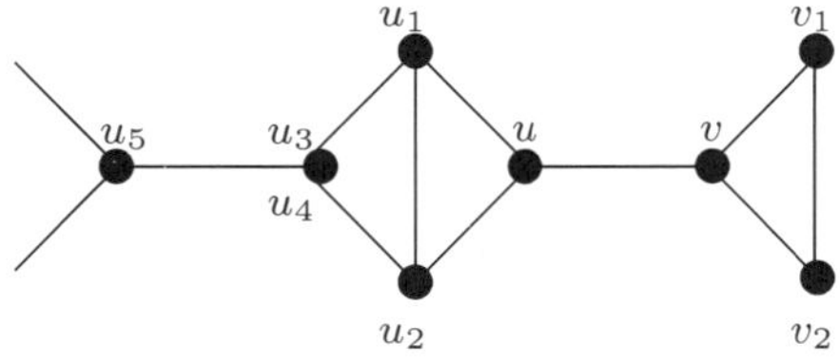

Fig. 10.18: Neighbour around the bridge (u, v) in G—Case 3a

Since $V(G_1) \cap V(G_2) = \emptyset$, (u_3, u_5) is also a bridge. Contracting G by removing the diamond set of vertices $\{u, u_1, u_2, u_3\}$ and connecting v to u_5 results in a bridge graph of order $N - 4$ which belongs to either Case 2 or Case 3, since $(v_1, v_2) \in E(G)$. Therefore, if we choose any cubic bridge graph of order $N - 4$ that belongs to either Case 2 or Case 3, insert a diamond set of four vertices and their associated edges like $\{u, u_1, u_2, u_3\}$ depicted in Figure 10.18 into the bridge, we obtain a cubic bridge graph of order N.

Note that, the set of cubic bridge graphs of order N obtained by applying such a construction to any cubic bridge graph in B_{N-4}^2 is equivalent to the set of cubic bridge graphs of order N obtained by applying the construction in Case 2a to any cubic bridge graph in B_{N-4}^2.

Therefore,

$$|B_N^{2a}| + |B_N^{3a}| = |B_{N-4}^1| + |B_{N-4}^2| + |B_{N-4}^3| = |B_{N-4}|. \tag{10.7}$$

Since the smallest possible cubic bridge graphs are of order 10, the smallest possible cubic bridge graph in Case 3a are of order 14.

Case 3b. Here, $u_3 \neq u_4$ (see Figure 10.19).

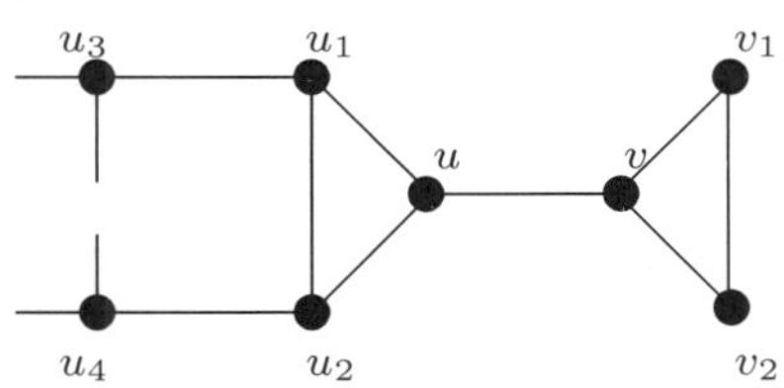

Fig. 10.19: Neighbour around the bridge (u, v) in G—Case 3b

Contracting G by removing the set of vertices $\{u_1, u_2\}$ and connecting u to u_3 and u_4 results in a bridge graph of order $N - 2$ which belongs to either Case 2 or Case 3, since $(v_1, v_2) \notin E(G)$. Therefore, if we choose any cubic bridge graph of order $N - 2$ that belongs to either Case 2 or Case 3, insert a set of two vertices and their associated edges like $\{u_1, u_2\}$ depicted in Figure 10.19 into one side of the bridge (the side that has a connecting edge between two vertices), we obtain a cubic bridge graph of order N.

Note that, the set of cubic bridge graphs of order N obtained by applying such a construction to any cubic bridge graph in B_{N-2}^2 is equivalent to the set of cubic bridge graphs of order N obtained from applying the construction in Case 2b to any cubic bridge graph in B_{N-2}^2.

Therefore,

$$|B_N^{2b}| + |B_N^{3b}| = |B_{N-2}^1| + |B_{N-2}^2| + |B_{N-2}^3| = |B_{N-2}|. \tag{10.8}$$

Since the smallest possible cubic bridge graphs are of order 10, the smallest possible cubic bridge graphs in this case are of order 12.

From (10.3),(10.7), and (10.8), we obtain

$$\begin{aligned}|B_N^1| + |B_N^{2a}| + |B_N^{3a}| + |B_N^{2b}| + |B_N^{3b}| &= |B_N^1| + |B_N^2| + |B_N^3| \\ &= \sum_i f(k_i, q_i) + |B_{N-4}| + |B_{N-2}|.\end{aligned}$$

□

In this chapter we outlined one attempt at counting cubic bridge graphs of a given order, and in the process introduced a new graph property, subdivision-equivalent edges. Arguably, the recursive relation described in Theorem 10.1 has potential to become an important component of a practical counting algorithm. However, to date, we do not have any tools for counting, or even estimating, the number of non-bridge non-Hamiltonian graphs of a given order. Thus, it is not yet clear whether Theorem 10.1 can be helpful in resolving Conjecture 9.1 concerning the prevalence of bridge graphs in the population of cubic non-Hamiltonian graphs. Perhaps, a fruitful approach to tackling this conjecture might be to explore a possibility of using results in this chapter together with newly proposed "genetic theory" of cubic graphs (see Baniasadi *et al.* [8]).

References

[1] G. Ahumada. Fonctions periodique et formule des traces de Selberg sur les arbres. *Comptes Rendus Mathematique Académie des Sciences*, 305:709–712, 1987.

[2] D. Aldous and J. Fill. *Reversible Markov chains and random walks on graphs.* (in preparation).

[3] N. Alon and J. H. Spencer. *The probabilistic method.* John Wiley & Sons, Inc., 2000.

[4] M. Andramonov, J. A. Filar, A. Rubinov, and P. Pardalos. *Hamiltonian cycle problem via Markov chains and min-type approaches*, pages 31–47. Approximation and complexity in numerical optimization: Continuous and discrete problems. Kluwer Academic, Dordrecht, 2000.

[5] K. Avrachenkov, J. Filar, and M. Haviv. *Singular perturbations of Markov chains and decision processes*, volume 40 of *International Series in Operations Research and Management Science*, pages 113–150. Kluwer Academic Publishers, 2002.

[6] K. E. Avrachenkov, M. Haviv, and P. G. Howlett. Inversion of analytic matrix functions that are singular at the origin. *SIAM Journal on Matrix Analysis and Applications*, 22(4):1175–1189, 2001.

[7] K. E. Avrachenkov and J. B. Lasserre. The fundamental matrix of singularly perturbed Markov chains. *Advances in Applied Probability*, 31(3):679–697, 1999.

[8] P. Baniasadi, V. Ejov, J. A. Filar, and M. Haythorpe. Genetic theory for cubic graphs. *Australasian Journal of Combinatorics*, submitted in 2011.

[9] R. B. Bapat and T. E. S. Raghavan. *Nonnegative matrices and applications.* Cambridge University Press, Cambridge, 1997.

[10] P. Berkhin. A survey on PageRank computing. *Internet Mathematics*, 2:73–120, 2005.

[11] M. Bianchini, M. Gori, and F. Scarselli. Inside PageRank. *ACM Transactions on Internet Technology*, 5:92128, 2005.

[12] D. Blackwell. Discrete dynamic programming. *Annals of Mathematical Statistics*, 33(2):719–726, 1962.

[13] B. Bollobas, T. Fenner, and A. M. Frieze. An algorithm for finding Hamiltonian paths and cycles in random graphs. *Combinatorica*, 7(4):327–341, 1987.

[14] J. A. Bondy and U. S. R. Murty. *Graph Theory with Applications.* New York: North Holland, 1976.

[15] V. S. Borkar. *Topics in controlled Markov chains.* Pitman Lecture Notes in Mathematics. Longman Scientific and Technical, Harlow, Essex, UK, 1991.

[16] V. S. Borkar. *Probability Theory: An advanced course.* Springer-Verlag, New York, 1995.

[17] V. S. Borkar, V. Ejov, and J. A. Filar. Directed graphs, Hamiltonicity and doubly stochastic matrices. *Random Structures and Algorithms*, 25:376–395, 2004.

[18] V. S. Borkar, V. Ejov, and J. A. Filar. On the Hamiltonicity gap and doubly stochastic matrices. *Random Structures and Algorithms*, 34(4):502–519, 2009.

[19] S. Brin and L. Page. The anatomy of a large-scale hypertextual Web search engine. *Computer Networks and ISDN Systems*, 33:107–117, 1998.

[20] A. Broder, A. M. Frieze, and E. Shamir. Finding hidden Hamiltonian cycles. In *Proceedings of the 23rd annual ACM Symposium on Theory of Computing*, volume 5, pages 395–410, 1994.

[21] F. Brunacci. Two useful tools for constructing Hamiltonian circuits. *European Journal of Operation Research*, 34:231–236, 1988.

[22] M. Chen and J. A. Filar. *Hamiltonian cycles, quadratic programming and ranking of extreme points*, pages 32–49. Recent advances in global optimization. Princeton Uiversity Press, Princeton, 1992.

[23] N. Christofides. *Graph theory: An algorithmic approach.* New York: Academic Press, 1975.

[24] F. Chung. *Spectral graph theory.* American Mathematical Society, 1997.

[25] E. Cinlar. Markov renewal theory: A survey. *Management Science*, 21(7):727–752, 1975.

[26] S. Cook. The P versus NP problem:
http://www.claymath.org/millennium/P_vs_NP/pvsnp.pdf
Last accessed: March 2012.

[27] D. M. Cvetkovic. *Graphs and their spectra.* PhD thesis, 1971.

[28] E. van Dam and W. Haemers. Which graphs are determined by their spectrum? *Linear Algebra and Its Application*, 373:241–272, 2003.

[29] J. Doob. *Stochastic processes.* Wiley, New York, 1953.

[30] V. Ejov, J. A. Filar, and J. Gondzio. An interior point heuristic for the Hamiltonian cycle problem via Markov decision processes. *Journal of Global Optimization*, 29(3):315–334, 2004.

[31] V. Ejov, J. A. Filar, M. Haythorpe, and G. T. Nguyen. Refined MDP-based branch-and-fix algorithm for the Hamiltonian cycle problem. *Mathematics of Operations Research*, 34(3):758–768, 2008.

[32] V. Ejov, J. A. Filar, S. K. Lucas, and J. L. Nelson. Solving the Hamiltonian cycle problem using symbolic determinants. *Taiwanese Journal of Mathematics*, 10:327–338, 2006.

[33] V. Ejov, J. A. Filar, S. K. Lucas, and P. Zograf. Clustering of spectra and fractals of regular graphs. *Journal of Mathematical Analysis and Applications*, 333:236–246, 2007.

[34] V. Ejov, J. A. Filar, W. Murray, and G. T. Nguyen. Determinants and longest cycles of graphs. *SIAM Journal on Discrete Mathematics*, 22(3):1215–1225, 2009.

[35] V. Ejov, J. A. Filar, and M. Nguyen. Hamiltonian cycles and singularly perturbed Markov chains. *Mathematics of Operations Research*, 19:223–237, 2004.

[36] V. Ejov, S. Friedland, and G. T. Nguyen. A note on the graph's resolvent and the multifilar structure. *Linear Algebra and Its Application*, 431(8):1367–1379, 2009.

[37] V. Ejov, N. Litvak, P. G. Taylor, and G. T. Nguyen. Proof of the Hamiltonicity-Trace conjecture for singularly perturbed Markov chains. *Journal of Applied Probability*, 48(4):901–910, 2011.

[38] V. Ejov and G. T. Nguyen. Consistent behavior of certain perturbed determinants induced by graphs. *Linear Algebra and Its Application*, 431(5-7):543–552, 2009.

[39] P Erdős. Some remarks on the theory of graphs. *Bulletin of the American Mathematical Society*, 53:292–294, 1947.

[40] P Erdős. Graph theory and probability. *Canadian Journal of Mathematics*, 11:34–38, 1959.

[41] A. Eshragh. *Hamiltonian cycles and the space of discounted occupational measures*. PhD thesis, 2010.

[42] A. Eshragh and J. A. Filar. Hamiltonian cycles, random walks and discounted occupational measures. *Mathematics of Operations Research*, 36(2):258–270, 2011.

[43] A. Eshragh, J. A. Filar, and M. Haythorpe. A hybrid simulation-optimization algorithm for the Hamiltonian cycle problem. *Annals of Operations Research*, 189:103–125, 2011.

[44] E. A. Feinberg. Constrained discounted Markov decision process and Hamiltonian cycles. *Mathematics of Operations Research*, 25(1):130–140, 2000.

[45] E. A. Feinberg and M. T. Curry. Generalized pinwheel problem. *Mathematical Methods of Operations Research*, 62:99–122, 2005.

[46] E. A. Feinberg and A. Shwartz. Constrained discounted dynamic programming. *Mathematics of Operations Research*, 21:922–945, 1996.

[47] J. A. Filar, A. Gupta, and S. K. Lucas. Connected co-spectral graphs are not necessarily both Hamiltonian. *The Australian Mathematical Society Gazette*, 32(3):193, 2005.

[48] J. A. Filar, M. Haythorpe, and G. T. Nguyen. A conjecture on the prevalence of cubic bridge graphs. *Discussiones Mathematicae Graph Theory*, 30(1), 2010.

[49] J. A. Filar and D. Krass. Hamiltonian cycles and Markov chains. *Mathematics of Operations Research*, 19:223–237, 1994.

[50] J. A. Filar and J.-B. Lasserre. A non-standard branch and bound method for the Hamiltonian cycle problem. *ANZIAM Journal*, 42(E):556–577, 2000.

[51] J. A. Filar and K. Liu. *Hamiltonian cycle problem and singularly perturbed decision process*, volume Statistics, probability and game theory: Papers in honor of David Blackwell: 30 of *IMS Lecture Notes-*

Monograph Series, pages 44–63. Institute of Mathematical Statistics, Hayward, CA, 1996.
[52] J. A. Filar and K. Vrieze. *Competitive Markov decision processes.* Springer, 1996.
[53] M. R. Garey, D. S. Johnson, and R. E. Tarjan. The planar Hamiltonian circuit problem is NP-complete. *SIAM Journal on Computing*, 5(4):704–714, 1976.
[54] B. Gavish and S. C. Graves. The Travelling salesman problem and related problems. Technical report, MIT, 1978.
[55] J. E. Gentle. *Random number generation and Monte Carlo methods.* Springer, 2nd edition, 2004.
[56] C. D. Godsil and B. D. McKay. Constructing cospectral graphs. *Aequationes Mathematicae*, 25:257–268, 1982.
[57] R. Greenlaw and R. Petreschi. Cubic graphs. *ACM Computing Surveys*, 27(4):471–495, 1995.
[58] J. Gross and J. Yellen. *Graph theory and its application.* CRC Press, 2005.
[59] F. Harary. *Graph theory.* Addison-Wesley, 1969.
[60] F. Harary and E. M. Palmer. *Graphical enumeration.* Academic Press, New York, 1973.
[61] T. H. Haveliwala and S. D. Kamvar. The second eigenvalue of the Google matrix. *Stanford University Technical Report*, 2003.
[62] M. Haythorpe. *Interior point and other algorithms for solving the Hamiltonian cycle problem.* PhD thesis, 2010.
[63] D. den Hertog. *Interior point approach to linear, quadratic and convex programming.* Kluwer Academic Publishers, 1994.
[64] D. Heyman. A decomposition theorem for infinite stochastic matrices. *Journal of Applied Probability*, 32:893–901, 1995.
[65] A. Hordijk and LCM Kallenberg. Constrained undiscounted stochastic dynamic programming. *Mathematics of Operations Research*, 9(2):276–289, 1984.
[66] R. A. Horn and C. R. Johnson. *Matrix analysis.* Cambridge University Press, 1990.
[67] J. J. Hunter. Mixing times with applications to perturbed Markov chains. *Linear Algebra and Its Application*, 417(1):108–123, 2006.
[68] Y. Ihara. On discrete subgroup of the two by two projective linear group over p-adic field. *Journal of the Mathematical Society of Japan*, 18(3):219–235, 1966.
[69] R. Karp. Probabilistic analysis of partitioning algorithms for the Travelling salesman problem in the plane. *Mathematics of Operations Research*, 2(3):209–224, 1977.
[70] J. G. Kemeny and J. Laurie Snell. *Finite Markov chains.* Springer-Verlag, 1976.
[71] W. Kocay and P.-C. Li. An algorithm for finding a long path in a graph. *Utilitas Mathematica*, 45:169–185, 1994.

[72] A. Sainte Lague. Les reseaux (ou graphes). *Memorial des sciences mathematiques*, 18, 1926.

[73] C. E. Langenhop. The Laurent expansion for a nearly singular matrix. *Linear Algebra and Its Application*, 4:329–340, 1971.

[74] A. N. Langville and C. D. Meyer. Deeper inside PageRank. *Internet Math.*, 1:335–380, 2003.

[75] J. Lauri. Pseudosimilarity in graphs – A survey. *Ars Combinatoria*, 46, 1997.

[76] E. L. Lawler, J. K. Lenstra, A. H. G. Rinooy Kan, and D. B. Shmoys. *The Traveling salesman problem: A guided tour of combinatorial optimization.* John Wiley and Sons, Chichester, 1985.

[77] L. Margolin. On the convergence of the Cross-Entropy method. *Annals of Operations Research*, 134:201–214, 2005.

[78] B. D. McKay. nauty: http://cs.anu.edu.au/~bdm/nauty/
Last accessed: March 2012.

[79] M. Meringer. Fast generation of regular graphs and construction of cages. *Journal of Graph Ttheory*, 30:137–146, 1999.

[80] P. Mnev. Discrete path integral approach to the Selberg trace formula for regular graphs. *Communications in Mathematical Physics*, 274:233–241, 2006.

[81] G. T. Nguyen. *Hamiltonian cycle problem, Markov decision processes and graph spectra.* PhD thesis, 2009.

[82] J. Nocedal and S. J. Wright. *Numerical optimization.* Springer Series in Operations Research. Springer, 1999.

[83] A. J. Orman and H. P. Williams. A survey of different interger programming formulations of the Travelling salesman problem. Technical report, London School of Economics and Political Science, 2004.

[84] I. Parberry. An efficient algorithm for the Knight's Tour problem. *Discrete Applied Mathematics*, 73:251–260, 1997.

[85] J. W. Pitman. Occupation measures for Markov chains. *Advances in Applied Probability*, 9:69–86, 1977.

[86] L. Posa. Hamiltonian circuits in random graphs. *Discrete Mathematics*, 14:359–364, 1976.

[87] M. L. Puterman. *Markov decision processes: Discrete stochastic dynamic programming.* Wiley-Interscience, 1994.

[88] R. C. Read. The enumeration of locally restricted graphs (ii). *Journal of the London Mathematical Society*, 35:344–351, 1960.

[89] R. C. Read. Some unusual enumeration problems. *Annals New York Academy of Sciences*, 175:314–326, 1970.

[90] R. Robinson. Counting cubic graphs. *Journal of Graph Theory*, 1:285–286, 1977.

[91] R. Robinson and N. Wormald. Almost all cubic graphs are Hamiltonian. *Random Structures and Algorithms*, 3(2):117–126, 1992.

[92] R. Robinson and N. Wormald. Almost all regular graphs are Hamiltonian. *Random Structures and Algorithms*, 5(2):363–374, 1994.

[93] K. W. Ross. Randomized and past-dependent policies for Markov decision processes with multiple constraints. *Operations Research*, 37(3):474–477, 1989.
[94] R. Y. Rubinstein. Optimization of computer simulation models with rare events. *European Journal of Operation Research*, 99:89–112, 1997.
[95] R. Y. Rubinstein and D. P. Kroese. *The Cross-Entropy method: A unified approach to combinatorial optimization, Monte-Carlo simulation and machine learning.* Springer-Verlag, 2004.
[96] S. Serra-Capizzano. Jordan canonical form of the Google matrix: A potential contribution to the PageRank computation. *SIAM Journal on Matrix Analysis and Applications*, 27(2):305–312, 2005.
[97] A. Sinclair. *Algorithms for random generation and counting: a Markov chain approach.* Birkhäuser, Boston-Basel-Berlin, 1993.
[98] TSPLIB. http://comopt.ifi.uni-heidelberg.de/software/TSPLIB95/ Last accessed: March 2012.
[99] A. Tucker. *Applied Combinatorics.* John Wiley & Sons, 1980.
[100] W. T. Tutte. *Recent progress in combinatorics.* Academic Press Inc, 1969.
[101] S. Vajda. *Mathematical programming.* Addison-Wesley, London, 1961.
[102] E. W. Weisstein. Dodecahedral Graph. From *MathWorld*—A Wolfram Web Resource.
http://mathworld.wolfram.com/DodecahedralGraph.html
Last accessed: March 2012.
[103] E. W. Weisstein. Horton Graph. From *MathWorld*—A Wolfram Web Resource.
http://mathworld.wolfram.com/HortonGraph.html
Last accessed: March 2012.
[104] N. Wormald. Enumeration of labelled graphs i: 3-connected graphs. *Journal of the London Mathematical Society*, 2(19):7–12, 1979.
[105] N. Wormald. Enumeration of labelled graphs ii: Cubic graphs with a given connectivity. *Journal of the London Mathematical Society*, s2-20(1):1–7, 1979.
[106] N. Wormald. On the number of automorphisms of a regular graph. *Proceedings of the American Mathematical Society*, 76(2):345–348, 1979.

Index

Printed by Printforce, the Netherlands